PROGRAMMED BEGINNING ALGEBRA

Copyright © 1963, 1971, by John Wiley & Sons, Inc.

All rights reserved. No part of this book may be reproduced by any means, nor transmitted, nor translated into a machine language without the written permission of the publisher.

Library of Congress Catalogue Card Number: 70-152495

ISBN 0-471-22301-8

Printed in the United States of America

10 9 8 7 6 5

PREFACE

Programmed Beginning Algebra consists of eight units of material, covering topics normally a part of a course in first-year algebra. The material is organized as follows:

Volume I	Unit I	Natural Numbers
	Unit II	Integers
Volume II	Unit III	First-Degree Equations and Inequalities in One Variable
	Unit IV	Products and Factors
Volume III	Unit V	Fractions and Fractional Equations
Volume IV	Unit VI	Graphs and Linear Systems
Volume V	Unit VII	Radicals
	Unit VIII	Quadratic Equations

This second edition of *Programmed Beginning Algebra* differs from the first edition in a number of ways, although the basic program, which was developed empirically over a long period of time and which has statistical reliability, has not been altered.

Numerous exercise sets have been added and appended to each unit. These exercise sets are optional reinforcements in the performance of routine algebraic processes, and they have been deliberately kept out of the program proper so as not to interfere with the basic learning sequence. Students may or may not use these exercise sets as is appropriate in each individual case. Page margins have been shaded to help the user locate exercise sets as well as other material in the appendix of each unit.

The physical size of each volume of the second edition has been reduced for convenient use without sacrificing the usability of the type. Also, shading has been introduced in the response column to help identify clarifying remarks that are not part of the response per se.

This program can be used in a variety of ways. For reasonably capable students who accomplish the entire program, it constitutes a complete first-year course in algebra. Preferably, the program should be supplemented with periodic classroom sessions or individual conferences with a qualified mathematics instructor; but, even without such assistance, a conscientious student can expect to achieve a degree of proficiency adequate to perform in a satisfactory manner on most standard achievement tests in beginning algebra.

Any one of the eight units can be used independently of the others. Each unit can be used as a self-tutor, to supplement and/or reinforce the same topic in any standard first-year algebra textbook. If classroom attendance is interrupted due to illness or for some other reason, the material missed in class can be covered by accomplishing the appropriate unit in this series. The statement of objectives and the table of contents at the beginning of each unit will give a good picture of the material covered herein.

The material, as presented, is somewhat traditional in point of view. However, set terminology and set symbolism are used where helpful, inequalities in one and two variables are discussed in conjunction with equations, solution sets of equations and their graphs are approached through the concept of ordered pairs, and emphasis is placed on the logical basis for the routine operations performed.

The units in the program are sequential, and, with the exception of Unit I, there are certain prerequisites necessary for each. Every volume, except Volume I, includes a prerequisite test, together with a suggested minimum score. This score is to be viewed as flexible. The primary purpose of the test is to provide the individual student with a means of determining whether or not he can expect to complete the material in the volume without undue difficulty.

IRVING DROOYAN
WILLIAM WOOTON

HOW TO USE THIS MATERIAL

The theory underlying programmed instruction is quite simple. The program presents you with information in small steps called frames, and, at the same time, requires you to make frequent and active written responses in the form of a word, a phrase, or a symbol. Programmed material is only as effective as you, the student, make it. In order to achieve the maximum benefit from this unit, you should follow the instructions carefully.

If you wish to measure what you learn, you should take one form of the self-evaluation tests at the end of each unit before beginning that unit. Then, upon completion, you can take the alternate form of the test and compare the two scores.

To progress through the program:

1. Place the response shield over the response column (the column on the left-hand side of the page) so that the entire column is covered.
2. Read the first frame on the right-hand side of the page carefully, noting, as you do, the place where you are asked to respond. Do not write a response until you have read the entire frame.
3. After deciding on the proper response, write it in the blank provided or at the end of the frame.
4. Slide the response shield down until it is level with the top of the next frame. This will uncover the correct response to the preceding frame and any additional remarks accompanying it.
5. Verify that you have made the correct response. You should almost always have done so, because the frames were constructed in such fashion that it will be difficult for you to make an error.
6. If you have made the correct response, repeat Steps 2 to 5 above with the next frame. If you have made an error on the frame, read the frame again, draw a line through your incorrect response and write the correct response. Following this, repeat Steps 2 to 5 above.

Essentially, this is all there is to your part of the task, and the program must assume the remainder of the responsibility for your learning. However, there are a few additional things you can do to improve the effectiveness of the material.

1. Do not try to go too fast. Each frame should be read carefully and you should think about the response you are going to make. Does it fit the wording of the frame? Does it make sense? In many frames you will find clues (called prompts) that are designed to aid you in making a correct response. Does your proposed response match the clue?

2. Mathematics is a language written in symbols. In order to understand mathematics, you have to make a conscious effort to read the symbolism as though it were written out in words.

3. Do not try to do too much at once. If you stay with your work too long at one time, you tend to become impatient, and the number of errors you make will probably rise. If you are making good progress, and are not tired, there is no reason to stop, but if you find yourself becoming bored, or find yourself making a high percentage of errors, take a break. Come back to the program later.

4. After taking a break, pick up your work a few frames behind where you stopped work. This will lead you back into the logical sequence with a minimum of confusion.

5. Do not be overly attentive to the remarks that accompany some of the responses in the response column. They are there to clear up points of confusion once in a while, but if you are not confused, don't waste time on them. The important thing as far as you are concerned is the response itself.

6. If you wish additional reinforcement with respect to certain parts of the program, you can complete the supplemental exercises that are provided at the end of each unit. Footnotes are used at appropriate places in the program to refer you to the exercises.

7. Keep note paper near at hand to enable you to solve problems that require several steps.

8. Complete the unit. If you do not complete the unit, you cannot expect to achieve the full benefit of the material.

UNIT III First-Degree Equations and Inequalities in One Variable

OBJECTIVES

Upon completion of the unit, the student should:

1. Be able to define and identify examples of "equation," "first-degree equation," "member of an equation," "solution or root of an equation," and "equivalent equations."

2. Be able to find solutions of some simple first-degree equations by inspection.

3. Know the meaning of the addition, multiplication, and division axioms for equations, and be able to apply the axioms to transform equations to equivalent equations.

4. Be able to solve first-degree equations that do not involve parentheses.

5. Be able to solve a first-degree equation in two variables for either variable in terms of the other.

6. Be able to solve simple word problems expressible in terms of first-degree equations.

7. Be able to solve simple first-degree inequalities.

8. Be able to graph the solution sets of first-degree equations and inequalities over the integers.

9. Be able to read and write the set notation $\{x \mid \text{condition on } x\}$.

CONTENTS

PREREQUISITE TEST

Complete this test and score your paper from the answers given at the end of the unit.

1. If x represents 2, then $2x + 1$ represents_____.
2. If y is equal to -3, then $1 - 4y$ is equal to_____.
3. If z represents 4, then $5z - z$ is equal to_____.
4. The sum of two negative numbers is a (positive/negative) number.
5. The sum of 3 and -7 is_____.
6. Write $6 - 7 - 8$ as a single numeral.
7. The product of a positive number and a negative number is a (positive/negative) number.
8. Simplify: $7 - 3y + 3y + 9$.
9. Multiply: $(-3)(-7)$.
10. Divide: $\frac{-15}{-3}$.

If you missed more than three problems on this test, you are probably not adequately prepared to start Unit III. We suggest that you prepare yourself by working through earlier units of Programmed Beginning Algebra.

Unit III

First-Degree Equations and Inequalities in One Variable

Remark. Finding solutions of what are called "equations" is one of the more important concerns of elementary algebra. This unit largely deals with ways of doing this. Before we start worrying about how to solve equations, however, we had better take a look at some vocabulary associated with equations.

equation

1. Any statement of equality between two expressions is called an equation. $2x + 3 = x + 1$ is an __________________ stating that the expression $2x + 3$ is equal to the expression $x + 1$.

equation

2. $x - 4 = 3$ is an __________________ stating that the expression $x - 4$ is equal to 3.

left

3. The expression to the left of the equals sign of an equation is called the left member of the equation, and that to the right is called the right member. Thus, in the equation $2x + 1 = 3$, $2x + 1$ is the_____ member.

7

4. The right member of the equation $3x + 2 = 7$ is_____.

equation; $2x - 4$

5. $2x - 4 = 3$ is an ________________ whose right member is 3 and whose left member is __________.

right member

6. $x = 4$ is an equation whose left member is x and whose __________ ________________ is 4.

solution

7. If the variable in an equation is replaced by a number, and the resulting statement is true, the number is called a solution of the equation. If x is replaced by 2 in the equation $x + 3 = 5$, the result is $2 + 3 = 5$, which is true. Therefore, 2 is a ________________ of the equation.

true; is

8. If x is replaced by 3 in the equation $x + 4 = 7$, the result is $3 + 4 = 7$, which is (true/false). Thus, 3 (is/is not) a solution of the equation.

root

9. Another name for "solution" is "root." The result of replacing x with 3 in the equation $10 - x = 7$ is $10 - 3 = 7$, which is true. Therefore, 3 is a solution or __________ of the equation.

false

10. The number 3 is not a solution of $3x + 2 = 12$, because $3(3) + 2 = 12$ or $9 + 2 = 12$ is a (true/false) statement.

is

11. The result of replacing y with 2 in $3y + 4 = 8 + y$ is $3(2) + 4 = 8 + (2)$ or $6 + 4 = 8 + 2$. Therefore, 2 (is/is not) a solution of the equation.

is

$\frac{15}{-5}$ does equal -3.

12. 15 (is/is not) a solution of $\frac{x}{-5} = -3$.

is

$\frac{5+3}{4}$ does equal 2.

13. 5 (is/is not) a solution of $\frac{x+3}{4} = 2$.

7

14. It can be seen by inspection that 5 is a solution of $x = 5$. Similarly, a solution of $x = 7$ is_____.

-3

-3 does equal -3.

15. A solution of $x = -3$ is____.

sum

16. An equation is a sentence written in symbols. The equation $x + 3 = 7$ can be read "The sum of x and 3 is 7." $x + 5 = 6$ can be read "The_______of x and 5 is 6."

12

17. $x + 10 = 12$ can be read "The sum of x and 10 is______.

3

Naturally, if 3 is added to 2, the result is 5.

18. In many simple cases, the solution of an equation is obvious from the way the equation reads. Thus, $x + 2 = 5$ is read "The sum of x and 2 is 5." Simply asking yourself "What must be added to 2 to yield 5?" suggests the obvious solution,____.

2

19. It can be seen by inspection that a solution of $x + 2 = 8$ is 6, because $6 + 2 = 8$. Similarly, a solution of $x + 3 = 5$ can be seen to be____.

3

Because $3 - 2 = 1$.

20. A solution of $x - 2 = 1$ is____.

9

21. A solution of $x - 3 = 6$ is____.

product

22. The equation $3x = 6$ can be read "The product of 3 and x is 6." The equation $5x = 10$ can be read "The______________of 5 and x is 10."

4

Of course. Isn't 12 the product of 3 and 4?

23. The solutions of many simple equations, such as $3x = 12$, are obvious from the way the equations read. Thus, $3x = 12$ is read "The product of 3 and x is 12." Asking yourself "What must be multiplied with 3 to yield 12?" suggests the obvious solution,____.

4; 4

24. A solution of $2x = 6$ can be seen by inspection to be 3, because $2(3) = 6$. A solution of $2x = 8$ can be seen to be_____, because 2(___) = 8.

2; 2

25. A solution of $3x = 6$ is____, because 3(___) = 6.

-3

26. A solution of $4x = -12$ is_______.

Remark. We have now learned what words such as "equation," "member of an equation," "solution," and "root" mean and, what is more, we can find solutions of some simple equations by simply reading the equations carefully and thinking about what they say. But all equations do not have obvious solutions. For example, the solution of $7 + x = 3 + \frac{4x}{5} + x$ is not so evident. In order to find solutions of equations such as this, we need some tools with which to work. To acquire these tools, we can begin by looking at what is meant by equivalent equations.

equivalent

This idea is important!

27. Equations such as $x + 1 = 5$ and $x = 4$ that have the same solution(s), in this case 4, are said to be equivalent equations. The solution of both $x + 2 = 5$ and $x = 3$ is 3. Therefore, these are ____________ equations.

equivalent

28. Because $2x = -2$ and $x = -1$ have the same solution, -1, the equations are __________________.

11

29. The equations $x - 5 = 6$ and $x = 11$ have the same solution, which is ______, and the equations are therefore equivalent.

11

30. If two equations are known to be equivalent, it is sometimes possible to find a solution of one by finding a solution of the other. Thus, if $3x - 4 = 2x + 7$ and $x = 11$ are equivalent equations, a solution of the first equation must be _____ because this is clearly a solution of the second.

4

31. If $2x - 1 = x + 3$ and $x = 4$ are equivalent, then a solution of the first equation is ____ because it is clearly a solution of the second equation.

-3

32. If $3x + 2 = x - 4$ and $x = -3$ are equivalent, a solution of both equations is _______.

-2

33. If $5x = 2x - 6$, $3x = -6$, and $x = -2$ are equivalent, a solution of both of the equations is _______.

Remark. Equivalent equations are equations having the same solutions. By this is meant that every solution of one equation is also a solution of the other and vice versa. Now how many solutions can we expect an equation to have? The answer to this question depends upon the type of equation. The equations we are concerned with here have only one solution.

first-degree

34. If a variable occurs only to the first power in an equation, the equation is called a first-degree equation. Because x occurs only to the first power in $2x + 3 = x - 1$, this is a ________ __________ equation.

first-degree

35. $3x + 2 = 7$ is a ________ __________ equation.

is not

36. $3x^2 + 2x - 1 = 0$ (is/is not) a first-degree equation.

One of the terms, $3x^2$, contains the variable to the second power.

is not

37. $2x = x^2 - 6$ (is/is not) a first-degree equation.

first

38. The expressions "first-degree equation" and "equation of the first degree" mean the same thing. Therefore, an equation such as $2x + 1 = 3$ is a first-degree equation or an equation of the ________ degree.

one; 7

39. In general, first-degree equations in one variable have only one solution. Thus, if one solution is found, the equation is said to be "solved." The linear equation $x = 7$ has only __________ solution. This solution is the number_____.

3

Which is evident from the equation $x = 3$.

40. The equation $3x + 2 = 2x + 5$ has only one solution. If this equation is equivalent to the equation $x = 3$, the solution of both equations is the number _____.

8

41. If $2x - x = 8$ is equivalent to $x = 8$, the solution of the first equation must be the number_____.

Remark. We have noted that first-degree equations have just one solution, and therefore, if we have two equivalent first-degree equations and if we can find the solution of one of them, we have found the solution of the other also, since they have the same solution. This is the whole story in solving equations. But how do we know when we have equivalent equations? The answer to this question depends upon where we obtain the two equations we are interested in. One way of obtaining an equivalent equation is by simplifying one that we already have.

$4x = 8$

42. Since $2x$ and x are like terms (the variable factors are identical), the equation $2x - x = 8$ can be written $x = 8$ by combining the terms in the left member. The equation $3x + x = 8$ can be written ______ $= 8$ by combining the like terms in the left member.

$3x = 12$

43. By combining the like terms in the left member and right member, respectively, the equation $4x - x = 9 + 3$ can be written _________.

5

44. Since $4x - 3x$ is another name for x, the equations $4x - 3x = 5$ and $x = 5$ are different ways of writing the same thing. That is, they are equivalent equations. The solution of both equations is_____.

$x = 6$

45. The like terms in the left member of $4x - 3x = 6$ can be combined to produce the equivalent equation ___________.

$x = 11$; 11

46. The like terms in each member of $9x - 8x = 4 + 7$ can be combined to form the equivalent equation ________. The solution of both of these equations is the number_____.

$x = 13$; 13

47. It is difficult to determine the solution of $12x - 11x = 18 - 5$ by inspection. However, after the like terms in each member have been combined to form the equivalent equation __________, the solution of both equations can be obtained by determining the solution from the second equation. The solution is____.

$x = 15$; 15

48. The solution of the equation $15x - 14x = 8 + 7$ can be obtained by forming the equivalent equation __________, whose solution is evident by inspection. The solution is_____.

-2

Did you first change the given equation to an equivalent equation, $x = -2$?

49. Solve $12x - 11x = -4 + 2$. The solution is_____.

3

50. Solve $5 - 2 = 9x - 8x$. The solution is_____.

-10

51. Solve $6x - 5x = -7 - 3$. The solution is_____.

Remark. Keep in mind what you are doing. You are trying to solve first-degree equations whose solutions are not obvious by finding equivalent equations whose solutions are obvious. You now know that one way to generate an equivalent equation is by combining any like terms in a given equation. But you must be able to do more than this if you are to be able to handle any and all first-degree equations. The following frames will show you another way of obtaining an equivalent equation from a given equation.

$7 + 5 = 7 + 5$

Or $12 = 12$.

52. Consider the equation $7 = 7$. The addition of 3 to each member produces the equation $7 + 3 = 7 + 3$ or $10 = 10$. The addition of 5 to each member of $7 = 7$ results in the equation ________________.

$15 + 8 = 15 + 8$

Or $23 = 23$.

53. Consider the equation $15 = 15$. The addition of 8 to each member produces the equation ____________________.

$9 - 3 = 9 - 3$

Or $6 = 6$.

54. The addition of -3 to each member of $9 = 9$ results in____________________.

five

55. In an equation such as $x - 3 = 5$, for any value of x for which the equation is true, $x - 3$ and 5 represent the same number, namely, f_______.

number

In this case each member represents the number fifteen.

56. If x is replaced by a solution of the equation, each member of $x - 8 = 15$ represents the same ____________________.

solution

57. If x is replaced by a____________________of the equation, each member of $x + 2 = 7$ represents the same number.

equivalent

Did you remember?

58. For any value of x that is a solution of the equation $x - 2 = 5$, each member represents the number 5. The addition of 2 to each member results in a new equation, $x - 2 + 2 = 5 + 2$ or $x = 7$, which is true for the same value of x as the original equation. That is, $x - 2 = 5$ and $x = 7$ are____________________ equations.

$x = 10$

59. If 4 is added to each member of $x - 4 = 6$, the result is the equivalent equation______________.

$x = 13$

60. If 3 is added to each member of $x - 3 = 10$, the result is the equivalent equation______________.

$6 = y$

$10 - 4$ is 6; $4 + y - 4$ is y.

61. If -4 is added to each member of the equation $10 = 4 + y$, the result is the equivalent equation $10 - 4 = 4 + y - 4$, or ________ $= y$.

$6 = x$

or $11 - 5 = 5 + x - 5$.

62. If -5 is added to each member of the equation $11 = 5 + x$, the result is the equivalent equation __________.

$0 = x$

$3 - 3$ is 0; $x + 3 - 3$ is x.

63. If -3 is added to each member of $3 = x + 3$, the result is________.

$x = 2$

64. The addition of -2 to each member of $x + 2 = 4$ results in the equivalent equation_______. The same result is obtained if 2 is subtracted from each member.

$x = -3$

65. The result of subtracting 7 from each member of $x + 7 = 4$ is the equation $x + 7 - 7 = 4 - 7$, which is equivalent to__________.

-3; 3

Given a choice, we prefer to add -3 to each member to obtain the equivalent equation $x = 11$.

66. The equation $x + 3 = 14$ can be transformed to the equation $x = 11$, either by adding_____to each member or by subtracting_____from each member.

adding

67. The equation $3x = 2x - 4$ is equivalent to the equation $x = -4$, because the second equation is obtained from the first by____________ $-2x$ to each member.

$-2x$

68. To transform the equation $3x = 2x + 4$ to the equivalent equation $x = 4$, it is necessary to add____ to each member of the first equation.

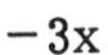

$-3x$

69. The equation $4x = 3x - 2$ can be transformed to the form $x = -2$ by adding _____ to each member.

axiom; equivalent

Recall that an axiom is simply a formal assumption.

70. The assumption

"If the same expression is added to each member of an equation, the result is an equivalent equation"

is called the addition axiom for equations. For example, if 4 is added to each member of $x - 4 = 3$, the addition ____________ guarantees that the resulting equation, $x = 7$, is ________________ to the original equation.

addition axiom

71. The addition axiom is used to transform an equation to an equivalent equation in which one member consists of just one term containing the variable. Thus, the addition of 4 to each member of $x - 4 = 7$ transforms this equation to $x = 11$. This transformation is justified by the

________________ __________.

5; $x = 3$

72. The equation $x - 5 = -2$ can be transformed to an equivalent equation in which x is the only term in the left member by adding ____ to each member. The resulting equation is ____________.

$2 = x$

73. An equation equivalent to $5 = x + 3$ can be obtained by adding -3 to each member. The equation is ____________.

2

74. Since $5 = x + 3$ and $2 = x$ are equivalent, the solution of both equations is the number _____.

-13

75. The equation $7 = x + 13$ can be transformed to the equivalent equation $7 - 13 = x + 13 - 13$ or $-6 = x$ by adding _______ to each member.

-6

76. Since $7 = x + 13$ and $-6 = x$ are equivalent, the solution of both equations is the number_____.

$x = 3$

> You may wish to follow the intermediate steps if you have not written the correct response.
>
> $$2x + 2 - 2 = x + 5 - 2$$
> $$2x = x + 3$$
> $$2x - x = x + 3 - x$$
> $$x = 3.$$

77. The equation $2x + 2 = x + 5$ can be transformed to an equivalent equation, in which the left member is x, by first adding -2 to each member and then adding $-x$ to each member and combining like terms. The resulting equation is__________.

$x = 3$

> $$2x + 2 = x + 5$$
> $$2x + 2 - x - 2 = x + 5 - x - 2$$
> $$x = 3.$$

78. By adding $-x - 2$ to each member, the equation $2x + 2 = x + 5$ can be transformed to the equivalent equation __________.

$-2x + 5$

> $5 - 2x$ is also correct. You could have thought of the process in two steps. Add $-2x$ or 5 to each member, then add the other term.

79. The equation $3x - 5 = 2x + 4$ can be transformed to the equivalent equation $x = 9$ by adding __________ to each member.

$-3x - 3$

> $-3 - 3x$ is also correct. Think of the process in two steps if you like. Add -3 or $-3x$ to each member, then add the other term.

80. The equation $4x + 3 = 3x - 7$ can be transformed to $x = -10$ by adding __________ to each member.

-10

81. The equation $4x + 3 = 3x - 7$ can be transformed to the equivalent equation $x = -10$ by applying the addition axiom. The solution of both equations is the number________.

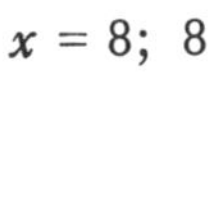

$x = 8$; 8

82. The two equations $3x - 3 = 2x + 5$ and $x = 8$ are equivalent. The equation with the most obvious solution is__________. The solution is the number_____.

$x = 10$; 10

83. Of the two equivalent equations $7x - 3 = 6x + 7$ and $x = 10$, the equation with the most obvious solution is__________. The solution is the number_____.

equivalent

84. In general, to solve a linear equation in one variable, it is helpful if an equivalent equation can be formed that has an obvious solution. It may be possible to apply the addition axiom to a given equation to produce an__________________ equation whose solution is obvious.

-17

85. The equation $x = -17$ has the obvious solution______.

6

86. The equation $6 = x$ has the obvious solution____.

$x = 5$

The solution, of course, is the number 5.

87. The solution of the equation $x + 2 = 7$ may be obvious to many. If the solution is not obvious, the addition of -2 to each member will produce the equivalent equation_________, whose solution is obvious.

$x = 8$; 8

88. The solution of the equation $3x - 2x = 8$ may be obvious to some. If the solution is not obvious, like terms in the left member can be combined to produce the equivalent equation_________, whose solution is obvious. The solution is the number

_____.

-6

89. The solution of a first-degree equation in one variable is most obvious if the variable appears by itself as one member of the equation. The solution of the equation $2x + 2 = x - 4$ is most obvious if the equation is transformed to the equivalent equation $x = -6$ by adding $-x - 2$ to each member. The solution of the original equation is the number______.

3; 15

90. The equation $12 = x - 3$ can be transformed to the equivalent equation $15 = x$ by adding____ to each member. The solution of $12 = x - 3$ is the number______.

$-2x$; 3

91. The equation $3 + 2x = 3x$ can be transformed to the equivalent equation $3 = x$ by adding_______ to each member. The solution of $3 + 2x = 3x$ is the number_____.

-2

-2 is clearly the solution of the equivalent equation $x = -2$.

92. The equation $7x + 5 = 3 + 6x$ can be transformed to the equivalent equation $x = -2$ by adding $-6x - 5$ to each member. The solution of $7x + 5 = 3 + 6x$ is the number_______.

adding

93. An equation whose solution may not be obvious can often be transformed to an equivalent equation whose solution is obvious by_______________ the same number to each member.

5

94. If the solution of the equation $19 = y + 14$ is obvious, write it directly.

If the solution is not obvious, find an equivalent equation whose solution is obvious and then write the solution.

-3

Adding $3 - 2y$ to each member of $3y - 3 = 2y - 6$ gives you $y = -3$.

95. If the solution of the equation $3y - 3 = 2y - 6$ is obvious, write it directly.

If the solution is not obvious, find an equivalent equation whose solution is obvious and then write the solution.

-5

96. Solve $6y + 2 = 5y - 3$. The solution is the number______.

14

Did you add $-11x + 5$ to each member?

97. Solve $12x - 5 = 11x + 9$. The solution is the number______.

12

If you had an incorrect solution, follow these steps.

$x - 2 = 10$

$x = 12$

98. Solve $4x - 2 - 3x = 4 + 6$ by first combining like terms and then adding 2 to each member. The solution is______.

7

99. Solve $3x - x = 7 + x$ by first combining like terms and then adding $-x$ to each member. The solution is_____.

3

100. Solve $4 = 3c - 2c + 1$. The solution is_____.

-4

101. Solve $0 = 3w + 5 - 2w - 1$. The solution is_____. *

* See Exercise 1, page 75, for additional practice.

Remark. This has been a long stretch of uninterrupted work, but the idea of using the addition axiom for equations in order to produce equivalent equations is extremely important, and deserves every bit of effort involved in learning it. Now, at the risk of telling you something you already know, let us review where we have been. We are interested in solving first-degree equations. If the equation is simple enough, we can find its solution (it has only one) by inspection. If the equation is too complicated for us to solve by inspection, we try to find an equivalent equation, that is, an equation with the same solution, whose solution we can determine by inspection. We have, so far, developed two ways of finding such equivalent equations, one by simply combining any like terms involved, and the other by applying the addition axiom for equations. Either or both of these ways can be used in solving an equation. This is where we now stand. The ways we have so far developed do not cover all of our needs. We shall next find another means for generating equivalent equations.

solution

"root" is also correct.

102. Like terms in a member of an equation may be combined to produce an equivalent equation that will have the same ________ as the original equation.

equivalent; solution

Again, "root" is correct, instead of solution.

103. If the same number is added to each member of an equation, the addition axiom guarantees that the result will be an equation that is ________ to the original equation. That is, the equations will have the same ________.

division

$$\frac{2x}{2} = \frac{6}{2}$$

$$x = 3.$$

104. Another useful axiom in solving equations is stated

> *"If each member of an equation is divided by the same nonzero number, the resulting equation is equivalent to the original equation."*

This is called the division axiom for equations. Thus, if each member of $2x = 6$ is divided by 2 to obtain the equivalent equation $x = 3$, the ________ axiom has been applied.

equivalent

105. If each member of $-6 = 3x$ is divided by 3, the division axiom for equations guarantees that the resulting equation, $-2 = x$, is ____________________ to the original equation $-6 = 3x$.

2

You probably knew the solution before applying the division axiom.

$$\frac{32x}{32} = \frac{64}{32}$$

$$x = 2.$$

106. The division axiom can be used to transform an equation such as $32x = 64$ to an equivalent equation, $x = 2$, by dividing each member by 32. The solution of both equations is_____.

$x = 3$

$$\frac{4x}{4} = \frac{12}{4}$$

$$x = 3.$$

107. The division axiom is used to transform equations such as $4x = 12$ to equivalent equations in which the left member is simply the variable x. If each member of $4x = 12$ is divided by 4, the result is the equation__________.

$x = 2$

108. If each member of $7x = 14$ is divided by 7, the result is the equation___________.

$7 = x$; 7

109. If each member of $35 = 5x$ is divided by 5, the result is the equation__________. Since this equation is equivalent to the original equation, the solution of both equations is_____.

$x = -3$; -3

110. If each member of $-4x = 12$ is divided by -4, the result is the equation ____________. The solution of both equations is_____.

Remark. The division axiom for equations assures you that you will get an equivalent equation if you divide each member by the same nonzero number. This is fine, providing you can choose the proper number for a divisor. This is not so difficult and we approach this problem by first reviewing the meaning of "coefficient."

coefficient

"Numerical coefficient" is also correct.

111. In speaking of the term $3x$, the symbol 3 is referred to as the numerical coefficient or simply the coefficient of the term. Thus 4 is the____________ of the term $4x$.

21

112. The coefficient of the term in the right member of $63 = 21x$ is_____.

1

113. If no numeral precedes a variable, for example, y, the coefficient of the term is understood to be 1. Thus, $y^3 = 1y^3$. The coefficient of the term x^2 is____.

coefficient

114. In the equation $3 = x$, the term in the right member has a ____________________ of 1.

1

115. The coefficient of the term in the left member of $x = -21$ is_____.

Remark. Now that we have reviewed the meaning of "coefficient," why do we need it?

1

116. The division axiom is used to find an equation with a coefficient of 1 on the term containing the variable. Using the division axiom and dividing each member by the coefficient of the term in the left member, $7x = 35$ is changed to an equivalent equation $x = 5$ where the coefficient of the term in the left member is_____.

-6; $x = -3$

$$\frac{-6x}{-6} = \frac{18}{-6}$$

$$x = -3.$$

117. To transform the equation $-6x = 18$ to an equivalent equation whose left member has a coefficient of 1, it is necessary to divide each member by______ the coefficient of $-6x$. The result is the equation ____________.

divide; $-2 = x$

118. To transform the equation $30 = -15x$ to an equation whose right member contains the variable with a coefficient of 1, it is necessary to__________ each member by -15. The result is the equation _________.

$3 = x$; 3

119. If each member of the equation $75 = 25x$ is divided by 25, the resulting equation is__________. The solution of both equations is______.

6

You probably did not have to apply the division axiom. If you did, however, you had

$$\frac{8x}{8} = \frac{48}{8}$$
$$x = 6$$

120. If the solution of the equation $8x = 48$ is obvious, write it directly. If the solution is not obvious, apply the division axiom to find an equivalent equation whose solution is obvious and then write the solution.

4

121. If the solution of the equation $7x = 28$ is obvious, write it directly. If it is not obvious, apply the division axiom to find an equivalent equation with an obvious solution and then write the solution.

9

You probably did not have to apply the division axiom.

122. The solution of $3x = 27$ is_____.

-4

$$\frac{-11x}{-11} = \frac{44}{-11}$$
$$x = -4$$

123. The solution of $-11x = 44$ is______.

-5

124. The solution of $5x = -25$ is______.

5

125. The solution of $-8x = -40$ is_____.

$-1y$

126. The coefficient of $-x$ is -1. Thus, $-x$ can be written $-1x$, and $-y$ can be written ______.

-6

$$\frac{-x}{-1} = \frac{6}{-1}$$
$$x = -6$$

127. The equation $-x = 6$ (or $-1x = 6$) can be transformed to the equivalent equation $x = -6$ by dividing each member by -1. The solution of both equations is______.

$4 = y$; 4

128. The equation $-4 = -y$ can be transformed to the equivalent equation ______ $= y$ by dividing each member by -1. The solution of both equations is_____.

-12

$$\frac{-x}{-1} = \frac{12}{-1}$$
$$x = -12.$$

129. The solution of $-x = 12$ is______.

6

130. The solution of $-6 = -y$ is_____.

Remark. At this point, you can generate equivalent equations by either combining like terms, applying the addition axiom, or applying the division axiom. Depending upon the first-degree equation you are to solve, you may need one or all of these procedures.

-4

$$\frac{2x}{2} = \frac{-8}{2}$$
$$x = -4.$$

131. It is frequently necessary to apply both the addition axiom and the division axiom to solve a first-degree equation. $3x + 2 = x - 6$ can be transformed to $2x = -8$ by adding $-x - 2$ to each member. The solution of each of these equations is______.

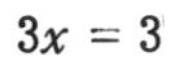

132. The addition axiom serves to eliminate terms. For example, the equation $4x - 5 = x - 2$ can be transformed by adding $-x + 5$ to each member, and the resulting equation, ___________, has fewer terms than the original equation.

adding

133. In using both the addition and division axioms, it is usually best to apply the addition axiom first. The equation $7x - 9 = 2x + 6$ could be transformed to $5x = 15$ by___________ $-2x + 9$ to each member.

dividing; 3

134. After transforming the equation $7x - 9 = 2x + 6$ to $5x = 15$ by means of the addition axiom, it is possible to transform this latter equation to $x = 3$ by ___________ each member by 5. The solution of all three equations is_____.

$9 = 3x$; $3 = x$

135. By adding $-2x + 4$ to each member of $2x + 5 = 5x - 4$, the equivalent equation___________ is obtained. This equation can then be transformed to __________by dividing each member by 3.

3

136. Since $2x + 5 = 5x - 4$ is equivalent to $3 = x$, the solution of both equations is_____.

$x = 6$

137. The addition axiom can be used to transform $7x - 3 = 5x + 9$ to the equivalent equation $2x = 12$. This equation can, in turn, be transformed by means of the division axiom to the equation $x =$ _____.

6

138. Since $7x - 3 = 5x + 9$ is equivalent to $x = 6$, the solution of both equations is_____.

2

Follow these steps if you have not written the correct answer.

$$3x + 4 - x - 4 = x + 8 - x - 4$$
$$2x = 4$$
$$x = 2.$$

139. The solution of $3x + 4 = x + 8$ is_____.

-2

140. The solution of $2x - 5 = 4x - 1$ is______.

-4

141. The solution of $3x + 2 = x - 6$ is______.

combining

"adding" is also correct.

142. If either or both members of an equation contain like terms, these terms should be combined before applying either the addition or division axiom. Thus, $3x + 2 - x = 6$ should be written $2x + 2 = 6$ by ____________________ the like terms in the left member before taking any other step in the solution.

2; 3

$$3x - 2 + 2 = 7 + 2$$
$$\frac{3x}{3} = \frac{9}{3}$$
$$x = 3.$$

143. After combining like terms in the left member, the equation $5x - 2 - 2x = 7$ has the form $3x - 2 = 7$. Adding______to each member and then dividing each member by______ results in the equivalent equation $x = 3$.

3

144. Since $5x - 2 - 2x = 7$ and $x = 3$ are equivalent equations, the solution of both is______.

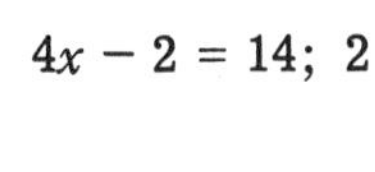

$4x - 2 = 14$; 2

145. By combining like terms, the equation $3x - 2 + x = 14$ can be written ____________________. The next step in the solution of the equation would be to add _____ to each member to obtain the equivalent equation $4x = 16$.

$x = 4$; 4

You could probably determine the solution of $4x = 16$ by inspection.

146. $3x - 2 + x = 14$ and $4x = 16$ are equivalent. The division axiom can be applied to the latter equation to produce another equivalent equation $x =$ _____, from which the solution of all three equations is clearly _____.

$x = 2$; 2

$$6x - 1 = 11$$
$$6x = 12$$
$$x = 2.$$

147. After (a) combining like terms, (b) applying the addition axiom, and (c) applying the division axiom, the equation $4x - 1 + 2x = 11$ is transformed to the equivalent equation $x =$ _____. The solution of the original equation is _____.

3

$$4x + 2 = 14$$
$$4x = 12$$
$$x = 3.$$

148. The solution of $3x + 2 + x = 14$ is _____.

6

149. The solution of $3x - 5 - x = 7$ is _____.

combining

Or "adding."

150. The equation $3x + 2 - x = 3x + 6 + x$ can be written $2x + 2 = 4x + 6$ by ________________ like terms.

addition

151. The equation $2x + 2 = 4x + 6$ can be transformed to $-4 = 2x$ by applying the ________________ axiom.

division

152. The equation $-4 = 2x$ can be transformed to $-2 = x$ by applying the ________________ axiom.

-2

153. The equations

$$\begin{aligned} 3x + 2 - x &= 3x + 6 + x, \\ 2x + 2 &= 4x + 6, \\ -4 &= 2x, \\ \text{and} \quad -2 &= x \end{aligned}$$

are equivalent. The solution of each of the equations is______.

-4

154. The solution of $2x + 1 - x = 3x + 5 - x$ is______.

$x + 1 = 2x + 5$
$-4 = x.$

2

155. The solution of $7x + 3 + 4 = 2x + 15 + 2$ is_____.

2

156. Solve $15x - 3 - 5x = 17 + x - 2$. The solution is _____.

-14

157. Solve $2 + 3x + x = 5x + 10 + 6$. The solution is ______. *

Remark. We now have established some formal machinery that can be used to help us find solutions of first-degree equations. The solutions of some equations can be determined by inspection, and wherever possible, this is obviously the most efficient procedure. However, in most cases, we need to find equations equivalent to given equations, because we cannot determine the solution of the given equation by inspection. We have found that we can generate equivalent equations by combining like terms or by applying the addition or division axioms. Now we need just one more basic tool, and we are going to be in a position to solve any first-degree equation.

x

158. The term $\frac{1}{3}x$ and the term $\frac{x}{3}$ are the same. Therefore, since $3\left(\frac{1}{3}x\right) = x$, $3\left(\frac{x}{3}\right) =$_____.

* See Exercise 2, page 75, for additonal practice.

y

159. The term $\frac{1}{5}y$ and the term $\frac{y}{5}$ are the same. Therefore, since $5\left(\frac{1}{5}y\right) = y$, $5\left(\frac{y}{5}\right) =$ _____.

$\frac{x}{9}$

160. The terms $\frac{1}{9}x$ and _____ are the same.

x

161. The product of 9 and $\frac{1}{9}x$, that is, $9\left(\frac{1}{9}x\right)$, equals _____.

x

162. The product of 9 and $\frac{x}{9}$, that is, $9\left(\frac{x}{9}\right)$, equals _____.

multiplication

163. A third axiom valuable in solving equations is stated:

> *"If each member of an equation is multiplied by the same nonzero number, the resulting equation is equivalent to the original equation."*

This is called the multiplication axiom for equations. Thus, if each member of $\frac{x}{5} = 2$ is multiplied by 5 to obtain the equivalent equation $5\left(\frac{x}{5}\right) = 5(2)$ or $x = 10$, the ____________________ axiom has been applied.

equivalent

164. If each member of $\frac{1}{3}x = 4$ is multiplied by 3, the resulting equation, $3\left(\frac{1}{3}x\right) = 3(4)$ or $x = 12$ is _______________ to the original equation, $\frac{1}{3}x = 4$.

21

165. The multiplication axiom can be used to transform an equation such as $\frac{x}{7} = 3$ to an equivalent equation $x = 21$ by multiplying each member by 7. The solution of both equations is_____.

3

166. The multiplication axiom is used to transform equations such as $\frac{1}{3}x = 6$ to equivalent equations in which there are no fractions. To transform $\frac{1}{3}x = 6$ to $x = 18$, it is necessary to multiply each member by ____.

2; 18

167. If each member of the equation $\frac{x}{2} = 9$ is multiplied by____, the resulting equation is $x = 18$, and the solution of both equations is_____.

3; −24

168. If each member of $\frac{x}{3} = -8$ is multiplied by____, the result is the equation $x = -24$. Since this equation is equivalent to the original equation, the solution of both equations is_____.

multiplied; 14

169. If each member of $7 = \frac{1}{2}y$ is_______________ by 2, the result is the equivalent equation $14 = y$. The solution of both equations is_____.

32

You probably didn't have to use the multiplication axiom to help you.

170. If the solution of $\frac{x}{2} = 16$ is evident by inspection, write it directly. If the solution is not evident, multiply each member by 2 and obtain an equivalent equation whose solution is obvious and then write the solution.

20

$(5)\frac{1}{5}x = 4(5)$

$x = 20.$

171. If the solution of $\frac{1}{5}x = 4$ is evident by inspection, write it directly. If the solution is not evident, apply the multiplication axiom to obtain an equivalent equation whose solution is obvious and then write the solution.

-49

172. If the solution of $-7 = \frac{1}{7}y$ is evident by inspection, write it directly. If the solution is not evident, apply the multiplication axiom to obtain an equivalent equation whose solution is evident and then write the solution.

30

173. The solution of $\frac{1}{6}x = 5$ is______.

24

174. The solution of $6 = \frac{x}{4}$ is______.

-21

175. The solution of $-3 = \frac{1}{7}x$ is______.

equivalent

176. The terms $\frac{3}{5}x$ and $\frac{3x}{5}$ are the same. Therefore, the equations $\frac{3}{5}x = 6$ and $\frac{3x}{5} = 6$ are eq______________t.

equivalent

177. $\frac{2}{7}x = 4$ and $\frac{2x}{7} = 4$ are________________equations.

$2x = 28$

178. If both members of $\frac{2}{7}x = 4$ are multiplied by 7, the resulting equation is________________.

14

179. $\frac{2}{7}x = 4$ and $2x = 28$ are equivalent equations. If the solution of the latter equation is evident by inspection, write it directly. If the solution is not evident, apply the division axiom and then write the solution.

12

This time the solution was probably not evident.
$3x = 36$
$x = 12$

180. If the solution of $\frac{3x}{4} = 9$ is evident by inspection, write it directly. If the solution is not evident, apply the multiplication axiom and the division axiom and then write the solution.

20

$40 = 2x$

$\frac{40}{2} = \frac{2x}{2}$

$20 = x$

181. If the solution of $8 = \frac{2}{5}x$ is evident by inspection, write it directly. If it is not evident, obtain equivalent equations until the solution is evident and then write the solution.

6

$30 = 5x$

$\frac{30}{5} = \frac{5x}{5}$

$6 = x$

182. Solve $10 = \frac{5x}{3}$. The solution is_____.

-14

183. Solve $-6 = \frac{3x}{7}$. The solution is______.

6

$2x = 12$
$x = 6.$

184. To solve $\frac{2x}{3} + 2 = 6$, it is advisable first to add -2 to each member and write the equivalent equation $\frac{2x}{3} = 4$. This equation can then be solved to find the solution to the original equation. The solution is_____.

$\frac{3x}{2} = 6$

185. To solve $\frac{3x}{2} + 7 = 13$, the addition axiom should be applied first. The addition of -7 to each member results in the equation ____ = ____.

$x = 4$

186. If both members of $\frac{3x}{2} = 6$ are multiplied by 2, the result is $3x = 12$. The division axiom can then be applied, and if each member is divided by 3, the result is ____ = ____.

4

You can easily determine this from the equation x = 4.

187. Because $\frac{3x}{2} + 7 = 13$, $\frac{3x}{2} = 6$, $3x = 12$, and $x = 4$ are all equivalent equations, the solution of $\frac{3x}{2} + 7 = 13$ is____.

6

$\frac{5x}{3} = 10$

$5x = 30$

$x = 6.$

188. Solve $\frac{5x}{3} - 3 = 7$ by applying the addition, multiplication, and division axioms in that order. The solution of the equation is____.

addition; multiplication; division

189. The equation $7 = 3 + \frac{4x}{5}$ can first be transformed to $4 = \frac{4x}{5}$ by means of the____________ axiom; this equation can then be transformed to $20 = 4x$ by means of the________________ axiom; and $20 = 4x$ can be transformed to $5 = x$ by means of the__________ axiom.

5

190. Since $7 = 3 + \frac{4x}{5}$ and $5 = x$ are equivalent equations, the solution of both is____.

7

$$6 = \frac{6x}{7}$$
$$42 = 6x$$
$$7 = x.$$

191. Solve $8 = 2 + \frac{6x}{7}$ by applying the addition, multiplication, and division axioms in that order. The solution is _____.

4

$$\frac{3y}{2} = 6$$
$$3y = 12$$
$$y = 4.$$

192. Solve $\frac{3y}{2} - 9 = -3$. The solution is____.

-20

193. Solve $19 + \frac{3x}{5} = 7$. The solution is_______.

Remark. This is all you need now to solve any first-degree equation. If you can combine like terms, and if you know how to apply the addition, division, and multiplication axioms for equations, that is all there is to it. Well, not quite all, you also have to know which axiom to apply and when to apply it.

addition; multiplication; division

194. Any first-degree equation can be solved by applying one or more of the three axioms for equations. These axioms are the a____________ axiom, the m__________________ axiom, and the d__________ axiom.

addition; left

You could have "same" instead of "left."

195. The addition axiom is used to transform an equation to an equivalent equation in which one or both members contain fewer terms. Thus, $3x + 4 = 7$ can be transformed to $3x = 3$ by applying the __________ axiom. The left member of $3x = 3$ has fewer terms than the________ member of $3x + 4 = 7$.

division; 1

196. The division axiom is used to transform an equation to an equivalent equation in which one member consists only of the variable with a coefficient of 1. Thus, $3x = 3$ can be transformed to $x = 1$ by applying the______________ axiom. The left member of $x = 1$ consists of the variable with a coefficient of_____.

multiplication

197. The multiplication axiom is used to transform an equation containing a fraction to an equivalent equation that contains no fraction. For example, $\frac{2x}{5} = 6$ can be transformed to $2x = 30$ by applying the ____________________ axiom.

addition

198. To transform $x + 3 = -9$ to $x = -12$, the ________________ axiom is applied.

division

199. To transform $3x = 15$ to $x = 5$, the ___________ axiom is applied.

multiplication

200. To transform $\frac{2x}{7} = 4$ to $2x = 28$, the ____________________ axiom is applied.

combine

"add" is also correct.

201. If more than one equivalent equation is necessary in the process of solving an equation, there is no specific order in which the axioms must be applied. In general, if like terms exist in the same member of an equation, it is advisable to combine them first. For example, to solve $\frac{5x}{3} - 6 = 11 + 3$, it is preferable to ______________ like terms first and write $\frac{5x}{3} - 6 = 14$.

adding

202. After combining like terms in the same member, the addition axiom can be applied to produce an equivalent equation in which all of the terms in one member contain the variable and all of the terms in the other member do not contain the variable. Thus $\frac{5x}{3} - 6 = 14$ can be transformed to $\frac{5x}{3} = 20$ by ___________ 6 to each member.

multiplying

203. Following the application of the addition axiom, if the term containing the variable appears in a fraction, the multiplication axiom can be applied. Thus, $\frac{5x}{3} = 20$ can be transformed to $5x = 60$ by ____________________each member by 3.

dividing

204. When necessary, the division axiom is applied to produce an equivalent equation in which one member of the equation consists of the variable with a coefficient of 1. Thus, $5x = 60$ can be transformed to $x = 12$ by_______________each member by 5.

solution

205. To solve linear equations the axioms for equations are applied to produce equivalent equations. This process is continued until an equation is obtained whose_______________is obtainable by inspection.*

Remark. You now not only have the necessary tools to solve first-degree equations, but also know which to use and when. To reinforce this knowledge, we shall take a look at some equations that contain more than one variable, and we shall apply our axioms to these equations in the same way we applied them to equations containing just one variable. Remember as you continue that the axioms to be used are those you already have learned, and that essentially they are applied for the same purposes. In any case where the division axiom is applied we shall assume that any variable involved does not take on the value 0.

dividing

206. The equation $d = rt$ can be transformed to the equivalent equation $\frac{d}{r} = t$ by_____________ each member by r.

* See Exercise 3, page 76, for additional practice.

t

207. The equation $d = rt$ can be transformed to the equivalent equation $\frac{d}{t} = r$ by dividing each member by_____.

solving

208. The process of transforming $d = rt$ to the equivalent equation $\frac{d}{r} = t$ is called "solving for t in terms of d and r" or, simply, "solving for t." Transforming $d = rt$ to the equivalent equation $\frac{d}{t} = r$ is called "______________ for r."

w

209. If $A = lw$ is transformed to $\frac{A}{l} = w$ by dividing each member by l, the equation is said to be "solved for_____."

$\frac{A}{w} = l$

210. To solve $A = lw$ for l, each member of the equation is divided by w, and the result is the equation

________.

divided; h

211. To solve $A = bh$ for b, each member of the equation is ____________ by_____.

$l = \frac{A}{w}$

212. In solving an equation for one variable in terms of another, it is customary to write the specified variable as the left member of the solved equation. Thus, in solving $A = lw$ for l, it is customary to write the result not as $\frac{A}{w} = l$ but as $l =$______.

symmetric

213. If $2 \cdot 4 = 8$, it is clear that $8 = 2 \cdot 4$. That is, if $a = b$, then $b = a$. This property is called the symmetric law of equality. Thus, in writing $l = \frac{A}{w}$ in place of $\frac{A}{w} = l$, the ________ law of equality is applied.

d

214. By the symmetric law of equality, if $d = rt$, then $rt =$ ____.

$r = \frac{d}{t}$

215. If $\frac{d}{t} = r$, then, by the symmetric law, ____ = ____.

symmetric

216. If $lwh = V$, then, by the ________ law, $V = lwh$.

dividing; t

217. An equation such as $s = vt$ can be transformed to $vt = s$ by means of the symmetric law; this equation can then be solved for v by ________ each member by ____, to obtain $v = \frac{s}{t}$.

$t = \frac{s}{v}$

$vt = s$

$t = \frac{s}{v}$.

218. Solve $s = vt$ for t by first applying the symmetric law and then applying the division axiom.

$t = \frac{s}{v}$

$\frac{s}{v} = t$

$t = \frac{s}{v}$.

219. Solve $s = vt$ for t by first applying the division axiom and then applying the symmetric law.

$h = \frac{V}{A}$

You could have used either the division axiom or the symmetric law first.

220. Solve $V = Ah$ for h.

$m = \frac{e}{c^2}$

221. Solve $e = mc^2$ for m.

$a = \frac{v}{t^2}$

222. Solve $v = at^2$ for a.

$180° - B - C$

223. To solve $A + B + C = 180°$ for A, the addition axiom is used. If $-B - C$ is added to each member, the result is $A =$ ________________.

addition

$-at^2$ is added to each member.

224. To solve $s = at^2 + b$ for b, the symmetric law is used to write $at^2 + b = s$, and then the ____________ axiom is used to write $b = s - at^2$.

$x = a - y$

225. To solve $x + y = a$ for x, the addition axiom is used. If $-y$ is added to each member, the result is ____________.

addition; $x = a$

226. To solve $x - a = 0$ for x, the ____________ axiom is applied. The result is ______.

$y = b - x$

227. Solve $x + y = b$ for y.

$-b$; a

228. The equation $ax + b = 0$ can be solved for x in terms of a and b by first adding _____ to each member to obtain $ax = -b$, and then dividing each member of this equation by ____ to obtain $x = \frac{-b}{a}$.

$x = \frac{-b}{2}$

$2x = -b$

$x = \frac{-b}{2}$.

229. Solve $2x + b = 0$ for x.

$x = \frac{-3}{a}$

$ax = -3$

$x = \frac{-3}{a}$.

230. Solve $ax + 3 = 0$ for x.

$\frac{c - d}{4}$

231. Recall that the quotient of $a + b$ divided by 3 is written $\frac{a + b}{3}$. The quotient of $c - d$ divided by 4 is written __________.

$\frac{2a - y}{3}$

232. The quotient of $2a - y$ divided by 3 is__________.

$x = \frac{c + d}{3}$

233. If both members of $2x = a - b$ are divided by 2, the result is $x = \frac{a - b}{2}$. If both members of $3x = c + d$ are divided by 3, the result is________________.

dividing; $x = \frac{y-3}{a}$

234. The equation $ax = y - 3$ can be solved for x by ______________ each member by a. The result is

______________.

$x = \frac{y+2}{b}$

235. Solve $bx = y + 2$ for x.

dividing; $x = \frac{3-y}{2}$

236. The equation $2x + y = 3$ can be solved for x by first applying the addition axiom to obtain $2x = 3 - y$ and then ______________ each member by 2 to obtain

______________.

$x = \frac{2+y}{3}$

$x = \frac{y+2}{3}$ is also correct.

237. To solve $3x - y = 2$ for x, the addition axiom is applied first to write the equivalent equation $3x = 2 + y$. Both members of this equation are then divided by 3 to obtain______________.

$x = \frac{7-y}{5}$

$5x = 7 - y$

$x = \frac{7-y}{5}$.

238. Solve $5x + y = 7$ for x.

$x = \frac{b+3}{a}$

$ax = b + 3$

$x = \frac{b+3}{a}$.

239. To solve $3 = ax - b$ for x, the symmetric law is first applied to write $ax - b = 3$. When solved for x, the final result is______________.

$x = \frac{b-3}{a}$

240. To solve $3 = b - ax$ for x, it is preferable to apply the addition axiom first and write $ax = b - 3$, because the sign associated with the ax term is then positive. When solved, the final result is______________.

$x = \dfrac{b + c}{a}$

$$ax - b = c$$
$$ax = b + c$$
$$x = \frac{b+c}{a}.$$

241. Solve $c = ax - b$ for x. Apply the symmetric law first.

multiplied

$$a\left(\frac{x}{a}\right) = a(b)$$
$$x = ab.$$

242. To solve the equation $\dfrac{x}{a} = b$ for x, the multiplication axiom is applied, and both members are ____________________ by a to obtain $x = ab$.

3; $x = 3c$

$$3\left(\frac{x}{3}\right) = 3(c)$$
$$x = 3c.$$

243. To solve $\dfrac{x}{3} = c$ for x, both members are multiplied by_____. The result is__________.

$x = 4b$

$$b\left(\frac{x}{b}\right) = b(4)$$
$$x = 4b.$$

244. Solve $\dfrac{x}{b} = 4$ for x.

$x = dr$

245. Solve $\dfrac{x}{d} = r$ for x.

divided

246. To solve $\dfrac{a}{b}x = 5$ for x, the multiplication axiom is applied to obtain $ax = 5b$. Both members of this equation are then______________by a to obtain $x = \dfrac{5b}{a}$.

$x = \frac{ab}{2}$

$a\left(\frac{2}{a}\right)x = a(b)$

$\left(\frac{a}{a}\right)2x = ab$

$2x = ab$

$x = \frac{ab}{2}$

247. To solve $\frac{2}{a}x = b$ for x, both members are first multiplied by a and then both members of the resulting equation are divided by 2. The result is _______.

$x = \frac{2c}{a}$

248. Solve $\frac{a}{2}x = c$ for x.

$x = \frac{ks}{r}$

249. Solve $\frac{r}{s}x = k$ for x.

$x = \frac{3b}{a}$

250. To solve $\frac{a}{b}x + 2 = 5$ for x, the addition axiom is applied first to write $\frac{a}{b}x = 3$. This equation is then solved for x to obtain _________.

$x = \frac{7c}{2}$

251. Solve $\frac{2}{c}x - 5 = 2$ for x.

$x = \frac{ac}{b}$

252. Solve $\frac{b}{c}x - a = 0$ for x.

$g = \frac{v - k}{t}$

253. To solve $v = k + gt$ for g, the symmetric law is applied first to write $k + gt = v$. By adding $-k$ to each member and then dividing each member by t, the resulting equation is ______________.

$t = \dfrac{v - k}{g}$

254. Solve $v = k + gt$ for t.

$a = \dfrac{s + 2t}{t^2}$

255. Solve $s = at^2 - 2t$ for a.

$r = \dfrac{A - P}{Pt}$

256. Solve $A = P + Prt$ for r.

$x = \dfrac{7 - y}{2}$

257. Solve $2x + y = 7$ for x.*

Remark. This is the last we shall say in this unit about solving first-degree equations. You have all of the techniques necessary, and you have applied the techniques in a variety of cases. We are next going to illustrate one of the ways in which equations are applied, namely, to help us find answers to questions about numbers as applied in the world about us. We shall only look at a few such examples, and these of a relatively simple nature, but it is quite important that you see how an equation can be used as a model for a real-life problem. We shall begin by examining some problems about numbers, and then go on to some examples involving numbers as measures of other things.

equation

258. A word sentence may be represented symbolically by an equation. For example, the sentence "A number added to 3 equals 7" can be represented symbolically by the________________ $x + 3 = 7$, where the variable represents a number.

$n + 4 = 12$

You might have $12 = n + 4$.

259. If the letter n is used as the variable, the sentence "A number added to 4 equals 12" may be represented by the equation________________.

* See Exercise 4, page 76, for additonal practice.

number

260. The sentence "Three times a number added to 6 equals 21" may be represented by $6 + 3n = 21$, where n represents a__________.

$2y - 6 = 24$

Of course, $24 = 2y - 6$ is also correct. We will not call this to your attention any more.

261. Use y to represent a number, and write the sentence "6 subtracted from two times a number equals 24" as an equation.

$n + 24 = 3n$

262. Let n represent a number and write the sentence "A number increased by 24 equals three times the number" as an equation.

$3x - x = 12$

263. The sentence "The difference between three times a number and the number itself is 12" can be represented by __________ $= 12$. Use x as the variable.

$5x - 3x = 8$

264. Write the sentence "The difference between five times a number and three times the same number is 8" as an equation. Use x as the variable.

$5x - 8$

265. The sentence in Frame 264 implies that three times the number is eight less than five times the number. This suggests the equation $3x =$__________.

$3x + 8$

266. The sentence in Frame 264 also implies that five times the number is eight more than three times the number. This suggests the equation $5x =$________.

$3x + 8$

267. The sentence in Frame 264 can be expressed by any of the three equations:

$$5x - 3x = 8,$$

$$3x = 5x - 8,$$

or $5x =$ ________.

$12 + 3x$; $7x - 12$

268. If x represents a number, the sentence "The difference between seven times a number and three times the same number equals 12" can be written in any one of the three forms:

$$7x - 3x = 12,$$

$7x =$ __________,

or $3x =$ __________.

$21 + x$; $8x - 21$

269. If x represents a number, the sentence "The difference between eight times a number and the number itself equals 21" can be written in any one of the three forms:

$$8x - x = 21,$$

$8x =$ __________,

or $x =$ __________.

9

270. The sentence "A number added to 6 equals 15" can be represented by $6 + n = 15$, where n is a number. The solution of the equation $6 + n = 15$ is _______.

is

271. To verify that 9 is the number identified by the sentence in Frame 270, the words "a number" can be replaced by 9 in the sentence, which then reads "9 added to 6 equals 15." Since the sentence is now true, 9 (is/is not) the number described.

$x + 5 = 9$

Any letter will do for the variable.

272. "Find a number such that five more than the number is 9." This sentence is what is called a "word problem" or a "stated problem." This condition on the number is expressed by the equation __________.

4

273. Solve $x + 5 = 9$.

is

274. Since five more than 4 is 9, 4 (is/is not) the solution of the word problem in Frame 272.

$4x + 1 = 29$

275. "If four times a certain number is increased by 1, the result is 29. What is the number?" Let x represent a number and write an equation representing the conditions on the number stated in the first sentence.

7

$4x + 1 = 29$
$4x = 28$
$x = 7.$

276. If the solution of $4x + 1 = 29$ is evident, write it directly. If the solution is not evident, transform the equation to an equivalent equation in which the solution is evident. The solution of $4x + 1 = 29$ is_____.

Yes

277. Does 7 satisfy the conditions of the problem in Frame 275?

$5y - 3 = 17$

278. "If five times a number is decreased by 3 the result is 17. Find the number." Let y represent a number and write an equation representing the condition on y.

4

279. The solution of $5y - 3 = 17$ is_____.

Yes

280. Does 4 satisfy the conditions of the problem in Frame 278?

Remark. Is it clear that we can use equations to identify numbers that have certain conditions imposed on them? We shall do some more of these, but before we do, we have to straighten out some facts about integers.

integer

281. Recall that the positive and negative whole numbers together with 0 form the set of integers. −4, −3, 0, 1, and 2 are examples of integers. 275 is an ______________.

integer

282. −275 is an ______________.

is

283. −7 (is/is not) an integer.

is not

284. 3/5 (is/is not) an integer.

consecutive

Or successive.

285. Consecutive or successive integers are integers that differ by 1. Thus, 3, 4, and 5 are ______________ integers, because they differ by 1.

9

286. The next consecutive integer after 8 is_____.

1

287. The next consecutive integer after 0 is_____.

−2

Remember, −3 is less than −2. Think of the number line in cases like this.

288. The next consecutive integer after −6 is −5 and the next consecutive integer after −3 is______.

1

Consecutive integers differ by 1

289. To go from one integer to the next consecutive integer,_____ is added to the given integer.

$x + 1$

Just add 1 to the given integer.

290. If x represents an integer, the next consecutive integer is represented by_________ .

$n + 1$

291. If n represents an integer, the next consecutive integer is represented by_________.

1

$n + 1 + 1 = n + 2$

292. The next consecutive integer to $n + 1$ is $n + 2$. This is because $n + 2$ is obtained from $n + 1$ by adding _____.

$x + 2$

293. The next two consecutive integers to x are $x + 1$ and_________.

$n + 1$; $n + 2$

294. Three consecutive integers can be written as n, _________ , and _________.

even integer

295. Even integers are integers that are exactly divisible by 2, therefore 8 is an______ _________.

even

296. Consecutive even integers differ by 2. The integers 4, 6, and 8 are examples of consecutive even integers. 12, 14, and 16 are also consecutive_________ integers.

$x + 2$

Simply add 2.

297. If x represents an even integer, the next consecutive even integer can be represented in terms of x by_________.

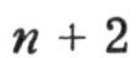

$n + 2$

298. If n represents an even integer, the next consecutive even integer is__________.

$x + 2$; $x + 4$

2 is added each time.

299. Let x represent an even integer, and write the next two consecutive even integers.

y; $y + 2$; $y + 4$

300. Let y represent the smallest of three consecutive even integers. Write the three integers.

odd integers

301. Odd integers are integers that leave a remainder of 1 when divided by 2. Another way of saying the same thing is that an odd integer is any integer that is not even. -7, 5, and 111 are_______ __________.

consecutive

302. Consecutive odd integers differ by 2; for example, 3, 5, 7, and 9. 13, 15, and 17 are also__________ odd integers.

$n + 2$; $n + 4$

2 is added each time.

303. Let n represent an odd integer and represent the next two consecutive odd integers.

y; $y + 1$; $y + 2$

You add 1 each time.

304. Let y represent the smallest of three consecutive integers. The integers can be represented by____, __________, and__________.

x; $x + 2$; $x + 4$

You add 2 each time.

305. If x represents the smallest of three consecutive even integers, the integers can be represented by ____, __________, and__________.

a; $a + 2$; $a + 4$

306. If a represents the smallest of three consecutive odd integers, represent the three integers in terms of a.

Remark. Now that you know some things about the properties of integers, you are ready to solve more word problems—this time about integers.

$n + 1$

307. "The sum of two consecutive integers is 37. Find the integers." If n represents an integer, the next consecutive integer is represented by__________.

$n + n + 1 = 37$

308. An equation expressing the conditions of the problem in Frame 307 is__________________________.

18

309. The solution to $n + n + 1 = 37$ is_______.

19

If $n = 18$, then $n + 1 = 19$.

310. The next consecutive integer after 18 is______.

Yes

311. Do 18 and 19 satisfy the conditions in the problem in Frame 307?

$n + 2$

312. "The sum of two consecutive even integers is 54. Find the integers." If n represents an even integer, the next consecutive even integer can be represented by__________.

$n + n + 2 = 54$

313. The equation representing the first sentence in Frame 312 is____________________.

26

314. The solution of $n + n + 2 = 54$ is______.

28

315. The next consecutive even integer after 26 is____.

Yes

316. Do 26 and 28 satisfy the conditions in the problem in Frame 312?

$n + 2$

317. "The sum of two consecutive odd integers is 36. Find the integers." Let n represent an odd integer, then__________represents the following odd integer.

$n + n + 2 = 36$

318. The equation representing the first sentence in Frame 317 is__________________.

17; 19

$$\begin{aligned} 2n + 2 &= 36 \\ 2n &= 34 \\ n &= 17 \\ n + 2 &= 19. \end{aligned}$$

319. Solve $n + n + 2 = 36$ and find the integers asked for in the word problem in Frame 317. These are_____ and_____.

23; 24

320. "The sum of two consecutive integers is 47." The integers are_____and_____.

-12; -13; -14

321. "The sum of three consecutive integers is -39." The integers are_____,_____, and_____.

47; 49

322. "The sum of two consecutive odd integers is 96." The integers are_____and_____.

-10; -12; -14

323. "The sum of three consecutive even integers is -36." The integers are______,______, and______.

Remark. Now, let us turn from word problems about numbers, to some word problems about measures of things; speeds, distances, lengths, and the like. To solve such problems, we have to know certain facts, usually facts in the form of formulas.

formula

324. An equation such as $A = lw$ that relates two or more variables is called a formula. This formula states that the area of a rectangle, in square units, is equal to the product of the length by the width. $A = \pi r^2$ is a_____________.

length

325. Formulas relate numbers. Thus, while $A = \pi r^2$ is spoken of as a relationship between the area of a circle and its radius, in reality, the formula relates the measure of the area and the length of the radius. When we say that a circle has a radius of 5 inches, we shall take it as understood that we mean the l__________ of the radius is 5 inches.

formula

326. Sometimes the solution of a word problem depends upon a formula which must be known in advance. For example, the distance (d) traveled by an object moving at a constant rate (r) for a given time (t) is given by the relationship $d = rt$, where d, r, and t represent numbers. $d = rt$ is a_____________.

$d = (45)(5)$

327. To solve the problem "How far does a car travel if it moves constantly at 45 miles per hour for 5 hours?", it is necessary to know the formula $d = rt$. Since the rate is 45 miles per hour and the time is 5 hours, the conditions given in the problem are represented by the equation $d = (\_\_\_)(5)$.

$24 = 4s$

Or $4s = 24$.

328. "How long is the side of a square whose perimeter is 24 feet?" To solve this word problem, the formula $P = 4s$ is used, where P represents the distance around the square (the perimeter) and s represents the length of each side. Since the perimeter is given as 24 feet, the conditions given in the problem are represented by the equation__________.

6

329. The solution of $24 = 4s$ is_____.

Yes

330. If each of the four sides of a square is 6 feet in length, is the perimeter 24 feet?

6 feet

The problem asks for a length.

331. The answer to the word problem in Frame 328 is ___ ________.

$3x$

332. "The perimeter of a rectangle is 56 feet. If its length is three times its width, what are the dimensions of the rectangle?" Let x represent the width of the rectangle. Then, in terms of x, ______ represents its length.

$x + 3x + x + 3x = 56$

A sketch is sometimes helpful to visualize the conditions.

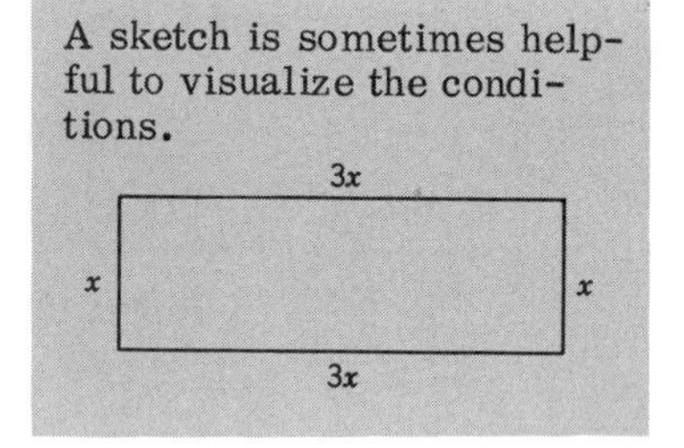

333. Since the perimeter of a rectangle is the sum of the length of its sides, an equation expressing the conditions of the problem in Frame 332 is ____________________________.

7

334. The solution of $x + 3x + x + 3x = 56$ is_____.

21

335. If x is equal to 7, then $3x$ is equal to______.

Yes

336. Does $7 + 21 + 7 + 21 = 56$?

7; 21

337. The width of the rectangle in Frame 332 is _____ feet, while its length is three times as great or _____ feet.

Remark. By now, the procedures involved in solving word problems should be fairly apparent. The last few problems suggest a profitable way to attack a word problem.

1. Read the problem carefully and represent the quantity or quantities to be found by means of symbols.
2. Where applicable, a sketch proves helpful, the sketch being labeled in terms of numbers and the variables being used.
3. Following this, the conditions expressed by the problem can be represented by means of an equation.
4. The equation can be solved and the solution of the equation can then be related to whatever it is the word problem asks.
5. Generally, as a last step, the answer is checked against the original word problem to insure that it meets the conditions that are set forth.

With this procedure in mind, let us look at some more examples.

symbol

338. "If 5 is added to twice a certain number, the result is 19. What is the number?" The steps suggested in the previous remark will be followed. The first step is to represent the number to be found as a s__________.

is not

339. Let n represent the desired number. Step 2 is to draw a sketch where applicable. In this problem a sketch (is/is not) applicable, since the conditions involve only numbers.

$2n + 5 = 19$

340. Write an equation representing the conditions stated in the problem in Frame 338. The equation is _____________.

7

341. Solve the equation. The solution of $2n + 5 = 19$ is_____.

Yes

342. If 5 is added to twice 7 is the result 19?

the same

343. Following the solution of the equation, the answer to the word problem should be stated. The solution of the equation in Frame 341 and the answer to the problem in Frame 338 are (the same/different).

$n + n + 2 = 86$

344. "The sum of two consecutive even integers is 86. Find the integers." Let n represent an even integer, then $n + 2$ represents the next consecutive even integer. The equation describing the condition on n is ____________________.

44

345. The solution of the equation $n + n + 2 = 86$ is 42. The two consecutive even integers referred to in Frame 344 are 42 and______.

Yes

346. Does the sum of 42 and 44 equal 86?

$n + 4$

347. "Find three consecutive odd integers whose sum is -33." If n represents an odd integer, then $n + 2$ and__________represent the next consecutive odd integers.

$n + n + 2 + n + 4 = -33$

348. The equation describing the condition on n is ______________________________.

-13

349. The solution of $n + n + 2 + n + 4 = -33$ is_____.

-9

350. If n is equal to -13, $n + 2$ is equal to -11, and $n + 4$ is equal to______.

Yes

351. Does the sum of -13, -11, and -9 equal -33?

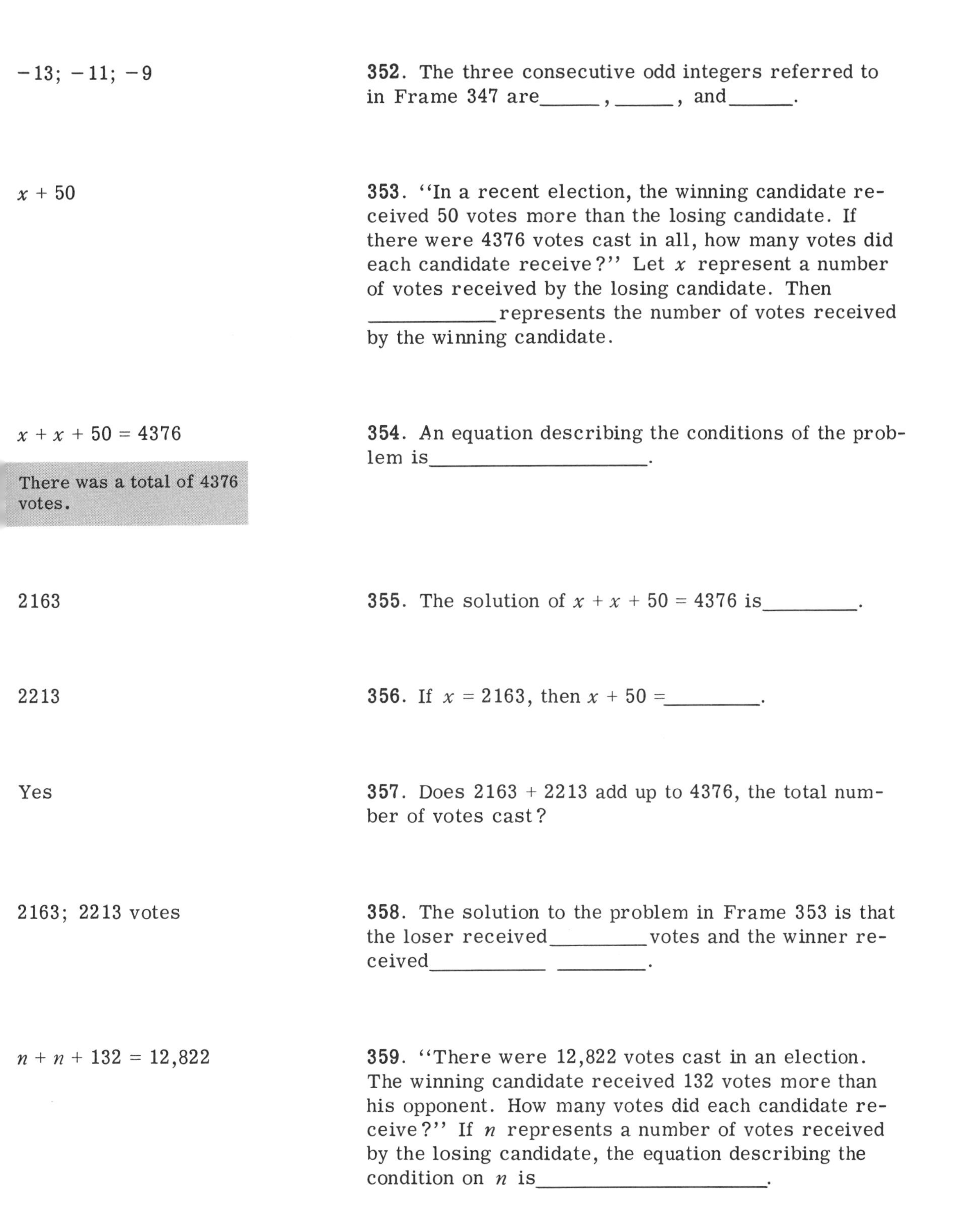

-13; -11; -9

352. The three consecutive odd integers referred to in Frame 347 are______, ______, and______.

$x + 50$

353. "In a recent election, the winning candidate received 50 votes more than the losing candidate. If there were 4376 votes cast in all, how many votes did each candidate receive?" Let x represent a number of votes received by the losing candidate. Then ____________ represents the number of votes received by the winning candidate.

$x + x + 50 = 4376$

There was a total of 4376 votes.

354. An equation describing the conditions of the problem is____________________.

2163

355. The solution of $x + x + 50 = 4376$ is_________.

2213

356. If $x = 2163$, then $x + 50 =$_________.

Yes

357. Does $2163 + 2213$ add up to 4376, the total number of votes cast?

2163; 2213 votes

358. The solution to the problem in Frame 353 is that the loser received_________votes and the winner received___________ _________.

$n + n + 132 = 12{,}822$

359. "There were 12,822 votes cast in an election. The winning candidate received 132 votes more than his opponent. How many votes did each candidate receive?" If n represents a number of votes received by the losing candidate, the equation describing the condition on n is____________________.

6345

360. The solution of $n + n + 132 = 12{,}822$ is_________.

6477

361. If $n = 6345$, then $n + 132$ is___________.

Yes

362. Does 6345 and 6477 add up to 12,822, the total votes cast?

6477 votes

363. The losing candidate received 6345 votes and the winning candidate received__________ __________.

$x + 3x = 24$

364. "A rope 24 feet long is cut into two pieces so that that one piece is three times as long as the other. How long is each piece?" If x represents a length of rope, $3x$ represents a length three times as long. A sketch would appear as

x | $3x$
24′

An equation describing the condition on x is
______________.

6

365. The solution of $x + 3x = 24$ is_____.

18

366. If $x = 6$, then $3x$ is______.

Yes

367. Does 6 feet and 18 feet equal 24 feet?

18 feet

368. The length of the shorter piece is 6 feet and the length of the longer piece is____ ________.

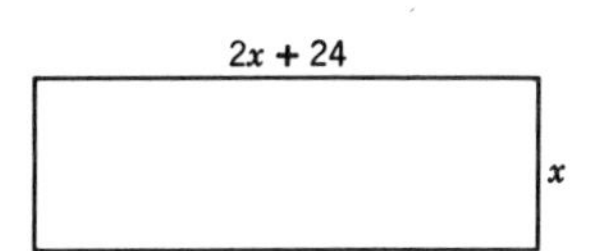

369. "A tennis court for singles is 24 feet longer than twice its width and its perimeter ($P = l+w+l+w$) is 210 feet. Find its dimensions." If x represents a width for the court, then $2x + 24$ represents its length, and a sketch would appear as

27

370. The equation describing the condition on x, determined by the formula

$$P = l + w + l + w,$$

is $2x + 24 + x + 2x + 24 + x = 210$. The solution of this equation is______.

78

371. If $x = 27$, then $2x + 24$ is______.

Yes

372. Does $27 + 78 + 27 + 78$ equal 210?

27; 78 feet

373. The width of the tennis court is______feet and the length (found by using $2x + 24$) is______ ______.

$x + 4$

374. "The perimeter of a triangle ($P = a + b + c$) is 56 inches. Two equal sides are each 4 inches longer than the base. Find the length of each side." If x represents a length for the base, then________represents the length of each of the equal sides.

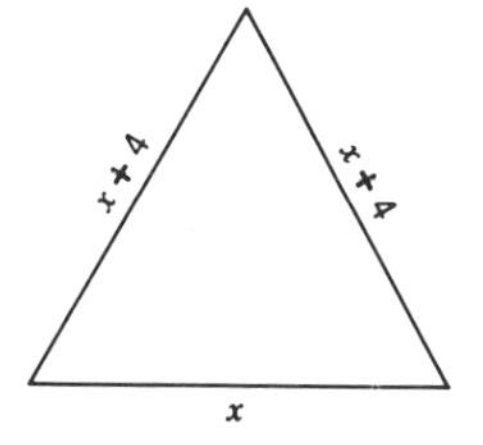

375. Sketch the triangle described in Frame 374 with sides and base labeled in terms of x.

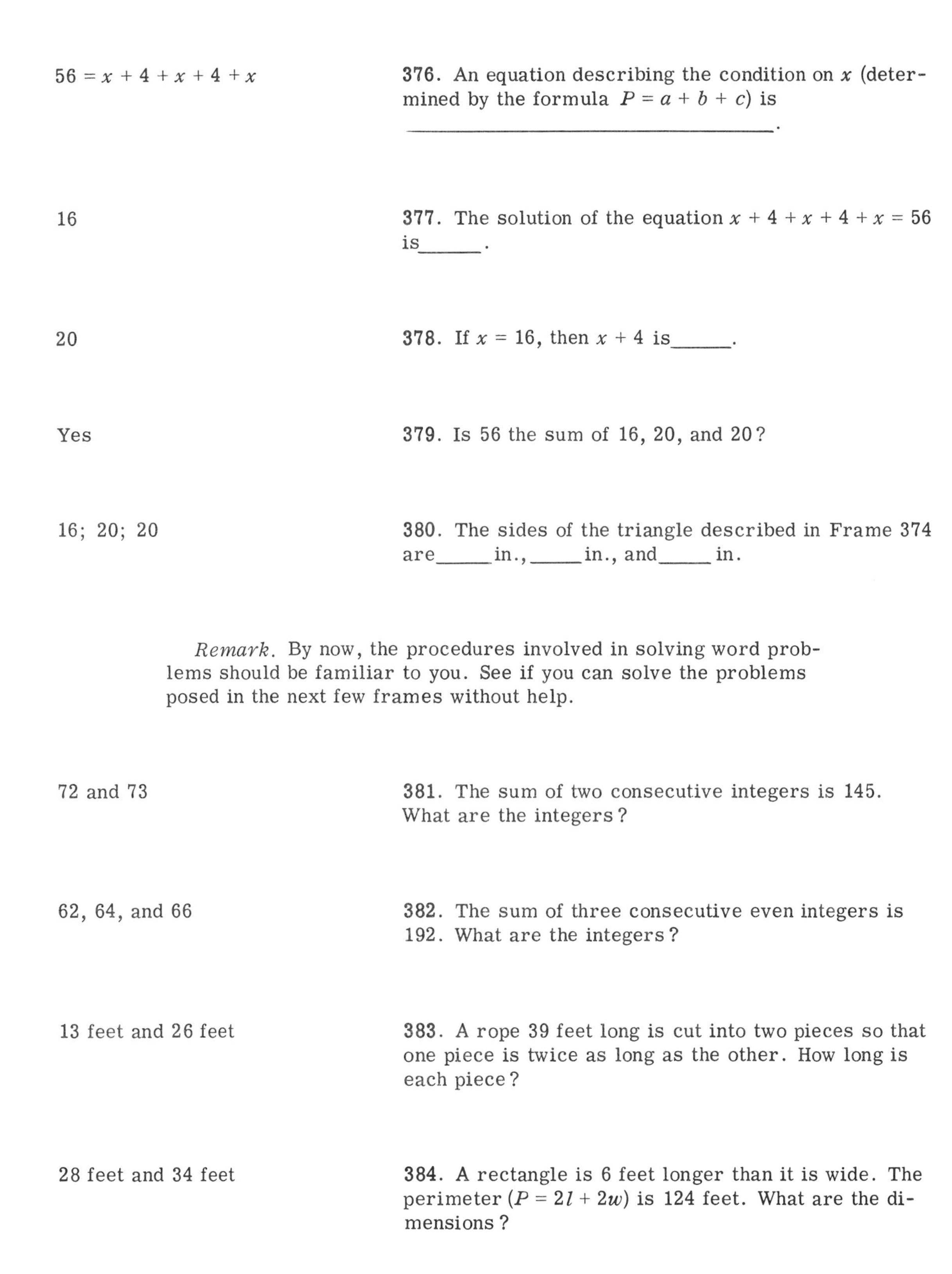

$56 = x + 4 + x + 4 + x$

376. An equation describing the condition on x (determined by the formula $P = a + b + c$) is __________________________________.

16

377. The solution of the equation $x + 4 + x + 4 + x = 56$ is______.

20

378. If $x = 16$, then $x + 4$ is______.

Yes

379. Is 56 the sum of 16, 20, and 20?

16; 20; 20

380. The sides of the triangle described in Frame 374 are_____in.,_____in., and_____in.

Remark. By now, the procedures involved in solving word problems should be familiar to you. See if you can solve the problems posed in the next few frames without help.

72 and 73

381. The sum of two consecutive integers is 145. What are the integers?

62, 64, and 66

382. The sum of three consecutive even integers is 192. What are the integers?

13 feet and 26 feet

383. A rope 39 feet long is cut into two pieces so that one piece is twice as long as the other. How long is each piece?

28 feet and 34 feet

384. A rectangle is 6 feet longer than it is wide. The perimeter ($P = 2l + 2w$) is 124 feet. What are the dimensions?

50°, 60°, and 70°

385. One angle of a triangle is 10° larger than another, and the third angle is 20° larger than the smallest. How large is each angle? Hint: The sum of the angles of a triangle equals 180°.*

Remark. This is all we propose to discuss with respect to first-degree equations at this time. There are, of course, many more types of applications of such equations, but the techniques involved in applying them are the same.

At this time, we propose to look at another type of relation, that involving the concept of inequality rather than equality. To begin with, let us review the meaning of some symbols.

less than

386. Recall that the symbol $<$ means "is less than." Thus, $3 < 5$ means "3 is ______ ______ 5."

4; 7

387. $4 < 7$ means "____is less than____."

$5 < 15$

388. "5 is less than 15" is written in symbols, ________.

$-7 < -3$

389. Write "-7 is less than -3" in symbols.

greater than

390. The symbol $>$ means "is greater than." Thus, $8 > 3$ means "8 is__________ __________ 3."

15; 11

391. $15 > 11$ means "_____ is greater than______."

$35 > 34$

392. "35 is greater than 34" is written in symbols, ______________.

* See Exercise 5, page 77, for additional practice.

$-8 > -10$

393. Write "-8 is greater than -10" in symbols.

left

394. On a line graph, of any two numbers, the point corresponding to the smaller number is located to the left of the point corresponding to the larger number. On the line graph

the point corresponding to -1 is found to the______ of the point corresponding to 2.

less

395. On the line graph

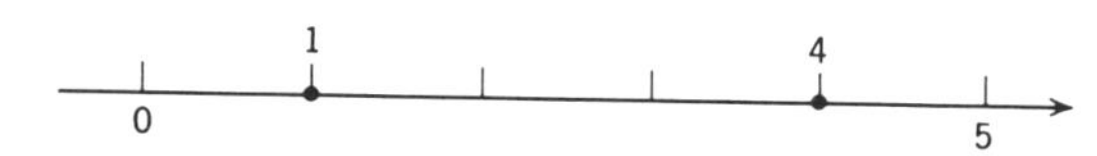

the point corresponding to 1 is to the left of the point corresponding to 4, and 1 is_________ than 4.

left; less

396. On the line graph

the point corresponding to the number a is to the _________of the point corresponding to the number b. Therefore, the number a is________ than the number b.

left

397. Because -7 is less than -1, the point corresponding to -7 on a line graph will be found to the _________ of the point corresponding to -1.

right

398. Because $8 > 5$, the point corresponding to the number 8 on a line graph will be found to the______ of the point corresponding to the number 5.

right

399. The point on a line graph corresponding to any number greater than 5 will be found to the ________ of the point corresponding to 5.

left

400. The point on a line graph corresponding to any number less than 7 will be found to the ________ of the point corresponding to 7.

Remark. We should now be ready to look at inequalities involving a variable. As with equations, our primary concern with these inequalities is going to be finding, and representing, solutions.

inequality

401. $x < 3$ is an example of a first-degree inequality in one variable. $x < -2$ is a first-degree ________________ in one variable.

variable

402. $x > -7$ is a first-degree inequality in one ________________.

first-degree

403. $x < 2$ is a ________________ inequality in one variable.

solution

404. If x is replaced with a number in $x < 2$, and the result is true, the number is called a solution of the inequality. When x is replaced by 1 in $x < 2$ the resulting statement $1 < 2$ is true. Thus, 1 is a ____________ of the inequality.

is

$3 < 5$ is true.

405. Because $2 < 7$, 2 is a solution of the inequality $x < 7$. 3 (is/is not) a solution of $x < 5$.

is

406. 5 (is/is not) a solution of $x < 10$.

is not

407. 7 (is/is not) a solution of $x < 5$.

$7 < 5$ is not true.

Remark. Inequalities differ from equalities in the number of solutions they have. Asking that one number be less than or greater than another is not as restrictive a condition as asking that one number be equal to another. Consequently, since the requirements are not as restrictive, we should expect to find more solutions for an inequality than for an equality. And so we do.

solution

408. The set of all solutions of an inequality is called the solution set of the inequality. Because 2 is less than 3, 2 is a member of the solution set of $x < 3$. 1 is also a member of the ________________ set of $x < 3$.

is

409. 3 (is/is not) a member of the solution set of $x < 10$.

$3 < 10$ is true.

is

410. -5 (is/is not) a member of the solution set of $x < 0$.

$-5 < 0$ is true.

is

411. Because the point corresponding to -7 is to the left of the point corresponding to -5, -7 (is/is not) a member of the solution set of $x < -5$.

$-7 < -5$ is true.

integer

412. Recall that the set of integers contains the natural numbers, their negatives, and zero. The set of integers can be represented in symbols $\{\ldots, -2, -1, 0, 1, 2, \ldots\}$, where the three dots indicate that the numbers continue in sequence indefinitely. Thus, -2 is an ________________ .

is

413. 5 (is/is not) an integer.

is

414. 0 (is/is not) an integer.

is not

415. 3/4 (is/is not) an integer.

is

416. −84 (is/is not) an integer.

is not

417. 2 1/2 (is/is not) an integer.

integers

418. Recall that a variable represents any number in a given set of numbers. The set of numbers is called the replacement set of the variable. If x is a variable that represents an integer, the replacement set of x is the set of ____________.

integer

419. If the replacement set of x is the set of integers, then, in any expression involving x, x can only be replaced by an ______________.

integer

420. If x represents an integer, then, in working with the inequality $x < 2$, we cannot replace x with 1/2, because 1/2 is not an ______________.

can

421. If x represents an integer, in $x < 3$, x (can/cannot) be replaced by −4.

422. $x = 7$ is an equation whose solution is the integer 7. The graph of the solution of $x = 7$ would appear as

where the point corresponding to 7 has been darkened and enlarged. Graph the solution of $x = 3$ on the line graph

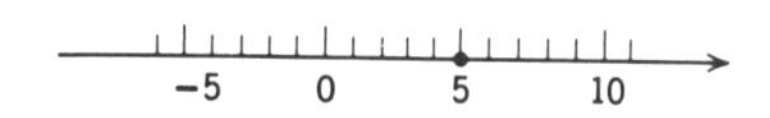

423. Graph the solution of $x + 3 = 8$ on the line graph

424. Graph the solution of $2x - 4 = -8$ on the line graph

infinite

425. Because there are an infinite number of numbers that are less than 5, the solution set of an inequality such as $x < 5$ contains an ____________ number of members.

Remark. Now here is a real problem. How do we go about showing an infinite number of solutions? There are two ways in which we propose to do it, the first one involves the use of the line graph. To help us keep the problem relatively simple, we are going to focus our attention only on the set of integers. For a while, we are going to be reminding you of this quite regularly. For the purposes of the work that follows, just pretend there are no numbers other than integers. In other words, you have never heard of numbers such as 1/2 or 3 3/4.

infinite

426. The solution set of an inequality can be shown be means of a line graph. If x is an integer, the solution set of $x < 5$ can be graphed as

where the three dots indicate that the graph continues indefinitely to the left. The solution set contains an ____________________ number of members.

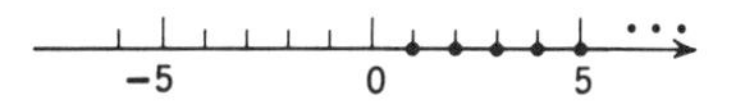

427. Since the point corresponding to any integer greater than 3 lies to the right of the point corresponding to 3, the graph of the solution set of $x > 3$, where x is an integer, appears as

where the three dots indicate that the graph continues indefinitely to the right. Graph the solution set of $x > 0$ (x is an integer).

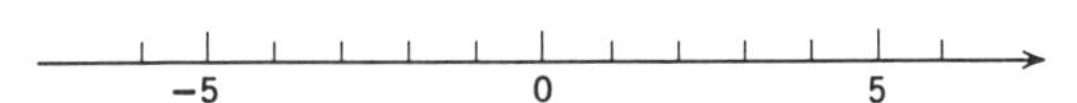

428. If x is an integer, the graph of the solution set of $x < 0$ appears as

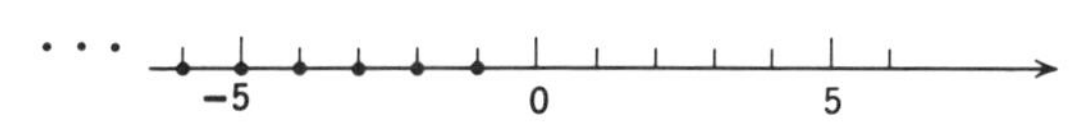

Graph the solution set of $x < -1$.

integers

429. Hereafter, when working with inequalities, it will be understood that the variables represent only elements of the set of integers. The solution set of $x > 5$ shall be the__________________greater than 5.

integer

430. In graphing solution sets of inequalities, the graph will consist only of points associated with the integers in the solution set. That is, even though 1/2 is a solution of $x < 1$, it would not be graphed because it is not an____________________.

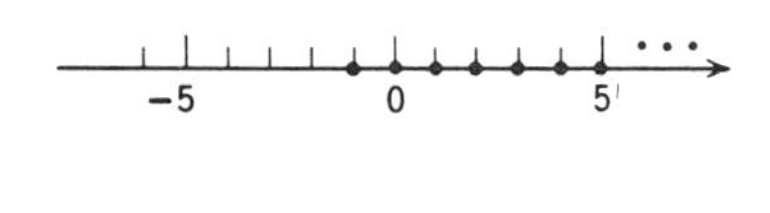

431. Graph the solution set of $x > -2$.

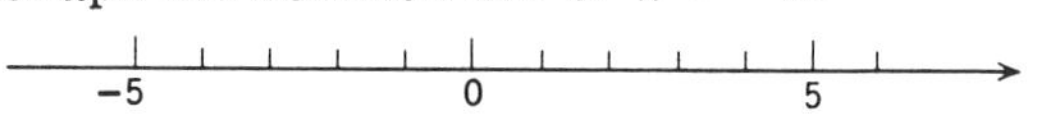

equal

432. The symbol $\leq$ means "less than or equal to." $x \leq 5$ means that x is less than or __________ to 5.

less; equal

433. $x \leq 0$ means that x is ________ than or ________ to 0.

greater

434. The symbol $\geq$ means "greater than or equal to." $x \geq 7$ means that x is ______________ than or equal to 7.

greater; equal

435. $x \geq -2$ means that x is _____________ than or _________ to -2.

is

436. The solution set of $x \leq 3$ differs from the solution set of $x < 3$ because, in the case of $x \leq 3$, 3 is in the solution set, while in the case of $x < 3$, 3 is not in the solution set. 4 (is/is not) in the solution set of $x \leq 4$.

is

437. 5 (is/is not) in the solution set of $x \geq 5$.

is not

$5 > 5$ is not true.

438. 5 (is/is not) in the solution set of $x > 5$.

is not

$8 > 8$ is not true.

439. 8 (is/is not) in the solution set of $x > 8$.

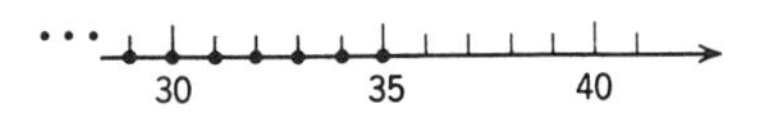

440. The graph of the solution set of $x \leq 15$ appears on a line graph as

The graph of $x \leq 35$ appears on a line graph as

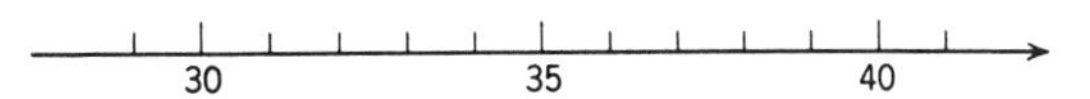

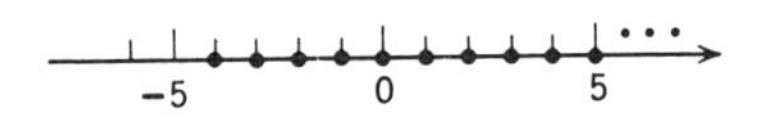

441. Graph the solution set of $x \geq -4$.

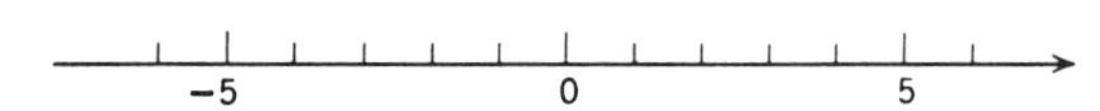

442. Graph the solution set of $x \leq 18$.

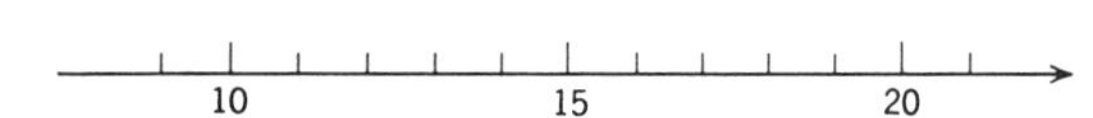

Remark. All of the inequalities discussed heretofore have involved only one term in each member, that is, have been of the type $x \leq 3$, or $x > 7$, or the like. However, just as in the case of equations, first-degree inequalities can involve more than one term in each member. While we aren't going into inequalities very deeply, we should at least take a quick look at how we might deal with inequalities such as $x + 3 < 5$.

addition

443. There are axioms dealing with inequalities that are very much like the axioms used to solve equations. For example, the addition axiom for inequalities is stated

> *If the same expression is added to each member of an inequality, the result is an equivalent inequality.*

For example, the fact that $x - 2 < 3$ is equivalent to $x - 2 + 2 < 3 + 2$ or $x < 5$ follows from the __________ axiom for inequalities.

-4

444. $x + 4 < 3$ can be transformed to $x < -1$ by adding _____ to each member.

$-x$

445. $2x < x + 5$ can be transformed to $x < 5$ by adding ______ to each member.

$x \geq 7$

446. If $-3x + 2$ is added to each member of $4x - 2 \geq 3x + 5$, the resulting equivalent inequality is ________________________ .

447. The addition axiom can be applied to some inequalities to produce equivalent inequalities whose solution sets are more obvious. Thus, the addition axiom can be applied to $3x < 2x + 3$ to produce $x < 3$, whose solution set can be graphed as

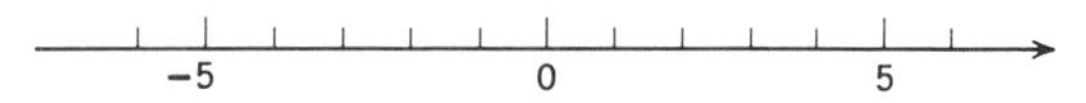

$x \geq -2$;

448. The solution set of $2x - 2 \geq x - 4$ is not obvious. However, if $-x + 2$ is added to each member of the inequality, the result is____________. The solution set can be graphed as

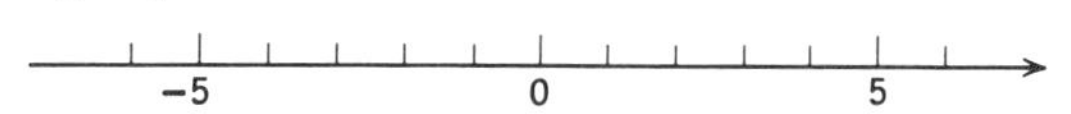

You first have $x \leq 2$.

449. By finding an equivalent inequality whose solution set is obvious, graph the solution set of $x + 7 \leq 9$.

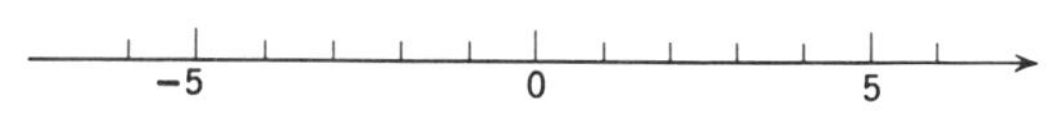

Did you first write $x > 5$?

450. Graph the solution set of $2x > x + 5$.

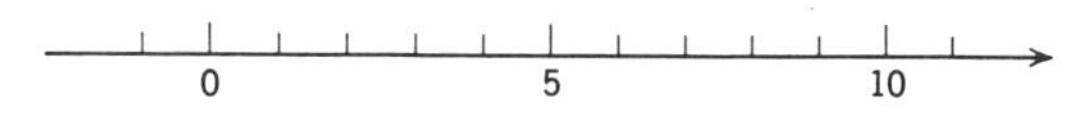

451. Graph the solution set of $3x - 4 \leq 2x - 5$.

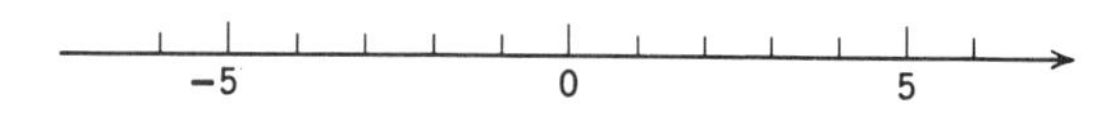

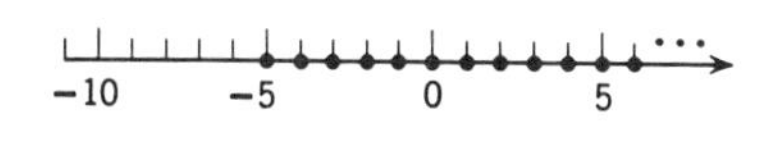

452. Graph the solution set of $5x + 3 \geq 4x - 2$.

greater

453. Observe that $x < 3$ can also be written $3 > x$. That is, if x is less than 3, then 3 is____________ than x.

$>$

"greater than" is also correct.

454. If $y < 7$, then 7____ y.

$\geq$

455. If $x + 2 \leq 2x$, then $2x$______ $x + 2$.

$2x - 1$

456. If $2x - 1 > 3x + 2$, then $3x + 2 <$____________.

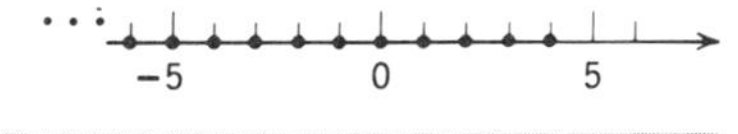

Did you first write the equivalent inequality $4 \geq x$?

457. The graph of the solution set of $-3 < x - 1$ is

and the graph of the solution set of $3 \geq x - 1$ is

458. Graph the solution set of $-6 > x - 2$.

459. Graph the solution set of $7 \geq x + 4$.

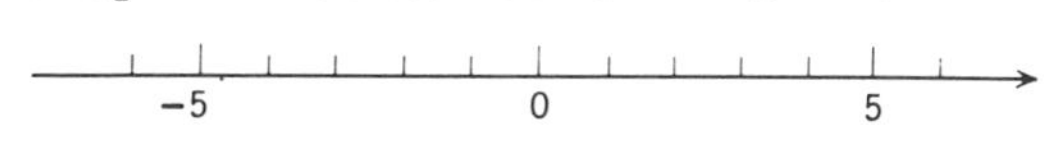

460. Graph the solution set of $3 < x - 1$

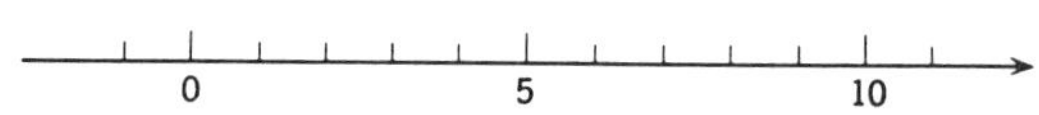

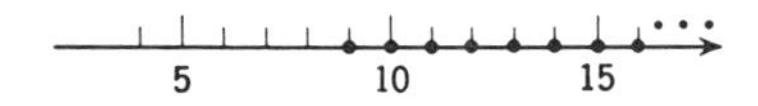

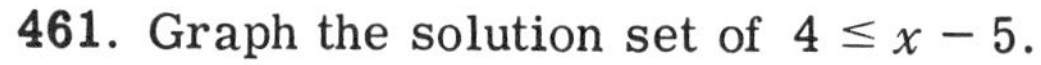

461. Graph the solution set of $4 \leq x - 5$.

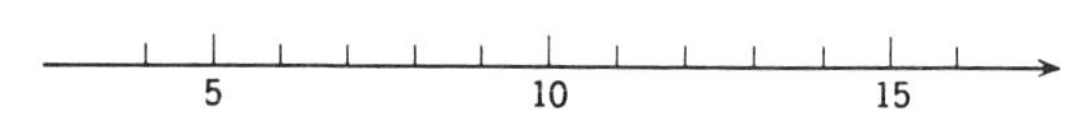

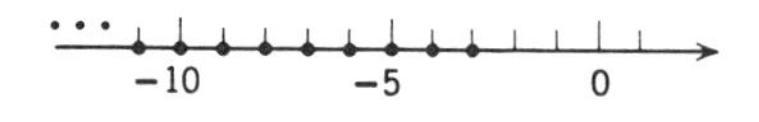

462. Graph the solution set of $4 \geq x + 7$

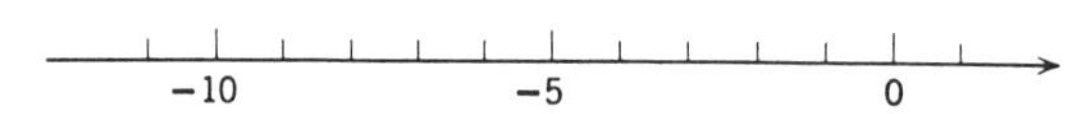

Remark. You will recall that we mentioned earlier that we would talk about two ways to show the solution set of an inequality. We have, at this point, only discussed the display of such sets on a line graph. Let us now look at another way to show them.

infinite

463. The graph of the solution set of $x > 3$ appears as

It is impossible to list the members in the solution set since there are a(n) (finite/infinite) number of members.

$x > 2$

464. A special set notation is used to describe a set with an infinite number of members. For example, the solution set of $x > 3$ is written as $\{x \mid x > 3\}$, which is read "the set of all x such that x is greater than three." Similarly, the solution set of $x > 2$ is written $\{x \mid$ _____ $\}$.

set

465. $\{x \mid x > 2\}$ contains an infinite number of members. In set notation, the braces, $\{\ \}$, refer to the set. $\{x \mid x < 4\}$ is read "the _______ of all x such that x is less than four."

such that

466. In set notation, the vertical bar, |, is read "such that." $\{y \mid y \geq 3\}$ is read "the set of all y ________________ y is greater than or equal to three."

$y < 4$

467. In the notation $\{y \mid y < 6\}$, $y < 6$ represents the condition on the variable y. The solution set of $y < 4$ is written $\{y \mid$ ________ $\}$.

greater than

468. $\{x \mid x > 5\}$ is read "the set of all x such that x is __________ ______ 5."

set; equal

469. $\{y \mid y \leq 3\}$ is read "the ______ of all y such that y is less than or __________ to 3."

such; less

470. $\{y \mid y \leq -4\}$ is read "the set of all y ______ that y is ________ than or equal to -4."

the set of all x such that x is greater than zero.

471. $\{x \mid x > 0\}$ is read "____."

$\{y \mid y \leq 0\}$

472. Represent "the set of all y such that y is less than or equal to zero" in set notation.

$\{x \mid x < 8\}$

473. Represent "the set of all x such that x is less than 8."

474. Graph $\{x \mid x > -1\}$

−5 0 5

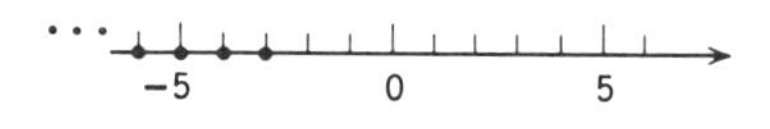

475. Graph $\{y \mid y \leq -3\}$ *

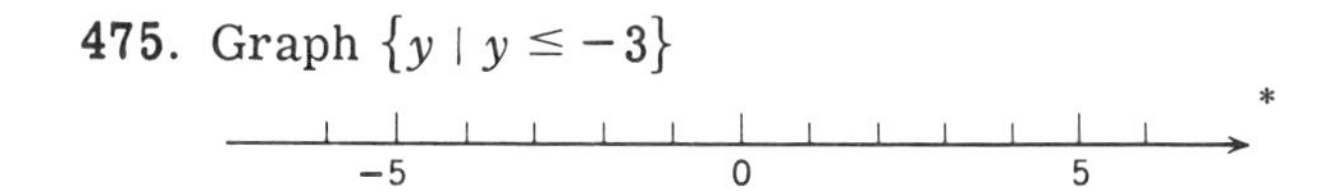

Remark. There are multiplication and division axioms that apply to inequalities, but they are a little different from those that apply to equations. You will encounter them in your next course in algebra, and we are not going to take them up here.

We have now covered all of the basic concepts involved in working with first-degree equations in one variable and all that we propose to cover with respect to first-degree inequalities. The following sequence of frames will give you an opportunity to obtain an overview of this unit in a very short time. Although you should not anticipate learning anything you have not learned to this point, the review will highlight the important ideas in a brief way and will reinforce what you have learned. Since these last few frames do not include a review of applications in the form of word problems, it might prove worthwhile at the end of this unit just to reread Frames 258 to 385.

equation

476. Any statement of equality between two expressions is called an equation. $4x + 3 = 2x - 4$ is an ________________.

solution

477. If the variable in an equation is replaced by a number, and the resulting equation is true, the number is called a solution or root of the equation. Since $3x + 1 = 7$ becomes $3(2) + 1 = 7$ when x is replaced by 2, 2 is a ________________ or root of the equation.

is not

$5 - 3 = 4$ is certainly not true.

478. 5 (is/is not) a solution of $x - 3 = 4$.

12

Too easy? The idea is that the solution of many equations can be obtained by inspection.

479. The solution of $x = 12$ is_____.

* See Exercise 6, page 78, for additional practice.

equivalent

480. Equations that have the same solution set are called equivalent equations. Because 12 is a solution of both, $2x - x = 11 + 1$ and $x = 12$ are ____________ equations.

combining

Or adding.

481. One way of transforming an equation such as $2x - x = 11 + 1$ to an equivalent equation $x = 12$ is by combining like terms in one or both members. $3x + x = 7 + 5$ can be transformed to the equivalent equation $4x = 12$ by ____________ like terms in each member.

addition

482. The addition axiom for equations asserts that the addition of the same number to each member of an equation produces an equivalent equation. Thus, transforming $3x + 2 = 2x + 7$ to the equivalent equation $x = 5$, is an application of the ____________ axiom for equations.

9

483. The solution of $3x - 1 = 8 + 2x$ is_____.

division

484. The division and multiplication axioms for equations assert that the division or multiplication of each member of an equation by the same non-zero number results in an equivalent equation. The equation $3x + 2 = 8$ can be transformed to the equivalent equation $x = 2$ by first applying the addition axiom to obtain $3x = 6$ and then applying the ____________ axiom to obtain $x = 2$.

2

The equation is equivalent to $x = 2$.

485. The solution of $3x + 2 = 8$ is_____.

addition; multiplication; division

486. The equation $\frac{2}{3}x + 4 = 16$ can be transformed to the equivalent equation $x = 18$ by first applying the ____________ axiom to obtain $\frac{2}{3}x = 12$, then applying the __________________ axiom to obtain $2x = 36$, and finally applying the __________ axiom to obtain $x = 18$.

18; equivalent

487. The solution of each of the equations

$$\frac{2}{3}x + 4 = 16$$

$$\frac{2}{3}x = 12$$

$$2x = 36$$

$$x = 18$$

is ____ because the equations are ____________.

40

You can, if necessary, follow these steps

$$\frac{3}{4}x = 30$$

$$3x = 120$$

$$x = 40$$

488. If the solution of $\frac{3}{4}x - 6 = 24$ is obvious by inspection, write it directly. If the solution is not obvious, find an equivalent equation whose solution is obvious and then write the solution.

dividing

489. An equation in two variables, such as $4x + 3y = 6$, can be transformed to an equivalent equation, $3y = 6 - 4x$, by adding $-4x$ to each member. The second equation can be transformed to $y = \frac{6 - 4x}{3}$ by __________ each member by 3.

$a = \dfrac{s - b}{t^2}$

You first have

$$at^2 + b = s$$
$$at^2 = s - b$$

and then $a = \dfrac{s - b}{t^2}$

490. Solve $s = at^2 + b$ for a. That is, obtain an equivalent equation with a the only variable in the left member.

less than

491. Statements of the form $x > 3$, $x \leq 4$, and $x + 2 < -6$ are called inequalities. $x < 3$ is an inequality stating that x is (less than/greater than) 3.

is not

492. 3, 4, and 5 are some solutions of $x > 2$ because $3 > 2$, $4 > 2$, and $5 > 2$ are true statements. 2 (is/is not) a solution of $x > 2$.

solution set

493. The set of all solutions of an inequality is called the____________ ________of the inequality.

494. The solution set of an inequality contains an infinite number of members. Such a solution set can be displayed by means of a line graph. If x is an integer, the solution set of $x > 2$ can be graphed as

where the three dots indicate that the graph continues indefinitely to the right. The graph of $x < 2$ would appear as

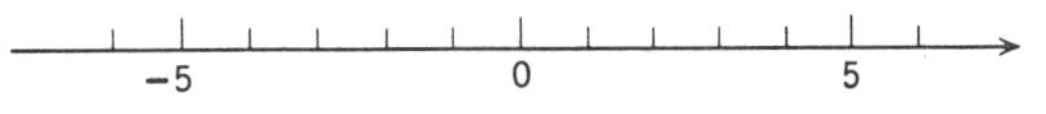

$\{x \mid x + 1 > 3\}$

495. Since the solution set of $x < 2$ contains an infinite number of members, it is necessary to use the notation $\{x \mid x < 2\}$ which is read "the set of all x such that $x < 2$" to denote the set. Similarly, the solution set of $x + 1 > 3$ can be represented by $\{x \mid$ __________$\}$.

set; x

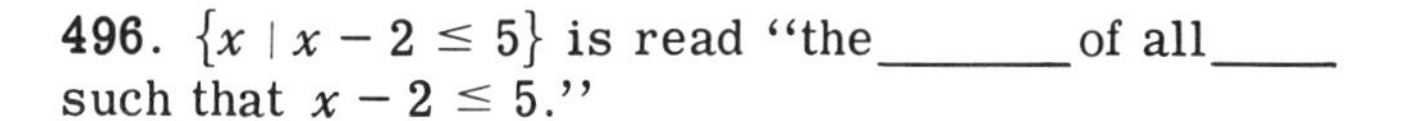

496. $\{x \mid x - 2 \leq 5\}$ is read "the_______ of all_____ such that $x - 2 \leq 5$."

497. Graph $\{x \mid x \geq -3\}$, where x is an integer.

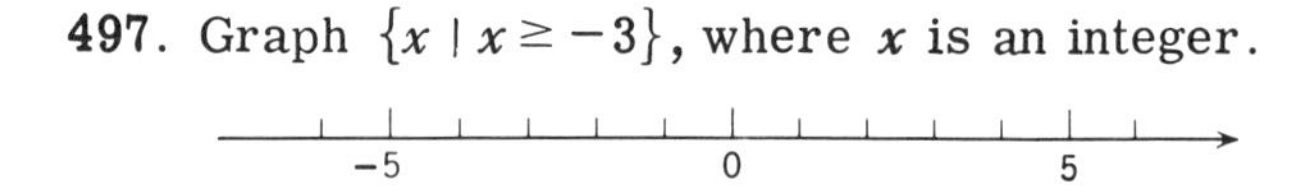

Remark. This concludes this unit. If you would like to see what you now know about first-degree equations and inequalities in one variable, you can take one form of the self-evaluation test on pages 80 to 81. If you completed one form of the test before you started the program, use the alternate form now.

EXERCISES AND ANSWERS

Exercise 1. If you have difficulty with this exercise, reenter the program at Frame 1.

Solve each equation.

1. $7x - 6x = 8$ **2.** $14x - 13x = 4$ **3.** $5x - 4x = 6 - 2$ **4.** $12x - 11x = 4 + 3$

5. $6 + 2 = 10x - 9x$ **6.** $8 - 10 = 15x - 14x$ **7.** $x + 12 = 14$ **8.** $x - 8 = 2$

9. $-3 = x + 4$ **10.** $-8 = x - 5$ **11.** $9x = 8x - 2$ **12.** $6x = 5x + 4$

13. $8x + 2 = 7x$ **14.** $5x - 7 = 4x$ **15.** $2x + 3 = x - 2$ **16.** $13x - 4 = 12x + 1$

17. $4x + 4 = 3x + 3$ **18.** $16x - 1 = 15x - 1$

Answers

1. 8 **2.** 4 **3.** 4 **4.** 7 **5.** 8 **6.** -2 **7.** 2 **8.** 10 **9.** -7 **10.** -3

11. -2 **12.** 4 **13.** -2 **14.** 7 **15.** -5 **16.** 5 **17.** -1 **18.** 0

Exercise 2. If you have difficulty with this exercise, reenter the program at Frame 102.

Solve each equation.

1. $8x = 24$ **2.** $-6x = 18$ **3.** $-12 = 4x$ **4.** $-35 = 5x$ **5.** $-2x = 8$

6. $-3x = -24$ **7.** $-x = 12$ **8.** $-x = -12$ **9.** $4x + 3 = x + 9$

10. $7x - 1 = 3x + 7$ **11.** $5x + 2 = 7x - 8$ **12.** $3x - 8 = 10x - 1$

Answers

1. 3 **2.** -3 **3.** -3 **4.** -7 **5.** -4 **6.** 8 **7.** -12 **8.** 12 **9.** 2

10. 2 **11.** 5 **12.** -1

Exercise 3. If you have difficulty with this exercise, reenter the program at Frame 158.

Solve each equation.

1. $\frac{x}{3} = 8$ 2. $\frac{x}{4} = -3$ 3. $4 = \frac{x}{2}$ 4. $-8 = \frac{x}{4}$ 5. $\frac{1}{5}x = 2$ 6. $\frac{1}{3}x = -6$

7. $\frac{2}{3}x = 8$ 8. $\frac{3}{5}x = -6$ 9. $4 = \frac{2}{7}x$ 10. $-12 = \frac{3}{4}x$ 11. $\frac{2}{3}x - 3 = 5$

12. $\frac{3}{5}x - 2 = 7$ 13. $\frac{3x}{4} + 4 = 1$ 14. $\frac{5x}{7} - 8 = -13$ 15. $9 = \frac{x}{7} + 2$

16. $16 = \frac{2}{3}x - 2$ 17. $-4 = \frac{5x}{6} + 1$ 18. $-12 = \frac{6x}{7} + 6$ 19. $10 = \frac{5x}{3} - 5$

20. $-4 = \frac{2x}{7} + 8$ 21. $\frac{3}{4}x - 2 = 7$ 22. $5 = \frac{4}{5}x - 7$ 23. $6 = \frac{3x}{2} + 9$

24. $-8 = \frac{4x}{7} - 8$

Answers

1. 24 2. −12 3. 8 4. −32 5. 10 6. −18 7. 12 8. −10 9. 14

10. −16 11. 12 12. 15 13. −4 14. −7 15. 49 16. 27 17. −6

18. −21 19. 9 20. −42 21. 12 22. 15 23. −2 24. 0

Exercise 4. If you have difficulty with this exercise, reenter the program at Frame 206.

Solve each equation for the specified variable.

1. $d = rt$ for t 2. $A = bh$ for h 3. $V = lwh$ for h 4. $A + B + C = 180°$ for C

5. $ay - b = 0$ for y 6. $b - ay = 0$ for y 7. $2x + 3y = 6$ for y 8. $2x - 3y = 6$ for x

9. $4 = 3x + 2y$ for y 10. $8 = 3x - 3y$ for x 11. $2xy - 5 = 12$ for y

12. $8 + 5xy = 4$ for x

Answers

1. $t = \frac{d}{r}$ 2. $h = \frac{A}{b}$ 3. $h = \frac{V}{lw}$ 4. $C = 180° - A - B$ 5. $y = \frac{b}{a}$ 6. $y = \frac{b}{a}$

7. $y = \frac{6 - 2x}{3}$ 8. $x = \frac{6 + 3y}{2}$ 9. $y = \frac{4 - 3x}{2}$ 10. $x = \frac{8 + 3y}{3}$ 11. $y = \frac{17}{2x}$

12. $x = \frac{-4}{5y}$

Exercise 5. If you have difficulty with this exercise, reenter the program at Frame 258.

a. Represent each word sentence by an equation. Use x as the variable. **b.** Solve the equation.

1. A number added to 5 equals 12.
2. Four times a number added to 3 equals 19.
3. Five times a number decreased by 7 equals 23.
4. Three times a number added to 4 equals 25.
5. The sum of two consecutive integers is 47.
6. The sum of two consecutive *even* integers is 66.
7. The sum of two consecutive *odd* integers is −84.
8. The sum of three consecutive integers is 87.
9. The sum of three consecutive *even* integers is 114.
10. The sum of three consecutive *odd* integers is −39.
11. The perimeter of a square is 36 feet. What is the length of a side?
12. The perimeter of a rectangle is 28 feet and its length is 4 more feet than its width. What are the dimensions?
13. The perimeter of a rectangle is 36 inches and its length is twice the length of the width. What are the dimensions?
14. There were 43,164 votes cast in an election. The winning candidate received 13,800 votes more than his opponent. How many votes did each candidate receive?
15. The perimeter of a triangle is 64 inches. Two equal sides are each 5 inches longer than the base. Find the length of each side?
16. One side of a triangle is 5 inches longer than another and the third side is 8 inches longer than the shortest side. What are the lengths of the sides if the perimeter is 64 inches?
17. One side of a triangle is 4 inches shorter than another side and the third side is 9 inches shorter than the longest side. What are the lengths of the sides if the perimeter is 44 inches?
18. One side of a triangle is 8 feet longer than another side and the third side is 4 feet shorter than the longest side. What are the lengths of the sides if the perimeter is 63 feet?
19. A square has the same perimeter as an equilateral triangle (3 equal sides). What is the length of each side of the triangle if the side of the square is 24 inches?
20. A rectangular garden is enclosed by 114 feet of a fence. If the width is 22 feet, what is the length?

Answers

1. a. $x + 5 = 12$ **b.** 7 **2. a.** $4x + 3 = 19$ **b.** 4 **3. a.** $5x - 7 = 23$ **b.** 6

4. a. $3x + 4 = 25$ **b.** 7 **5. a.** $x + x + 1 = 47$ **b.** 23, 24 **6. a.** $x + x + 2 = 66$ **b.** 32, 34

7. a. $x + x + 2 = -84$ **b.** $-43, -41$ **8. a.** $x + x + 1 + x + 2 = 87$ **b.** 28, 29, 30

9. a. $x + x + 2 + x + 4 = 114$ **b.** 36, 38, 40

10. a. $x + x + 2 + x + 4 = -39$ **b.** $-15, -13, -11$

11. 9 feet **12.** 5′ × 9′ **13.** 6″ × 12″ **14.** 28, 482, and 14,682

15. 18, 23, and 23 inches **16.** 17, 22, and 25 inches **17.** 19, 15, and 10 inches

18. 17, 25, and 21 feet **19.** 32 inches **20.** 35 feet

Exercise 6. If you have difficulty with this exercise, reenter the program at Frame 386.

Graph the solution set of each inequality. Assume that x represents an integer.

1. $x > 5$ **2.** $x < 5$ **3.** $x > -3$ **4.** $x < -3$ **5.** $x \leqslant 2$ **6.** $x \geqslant -1$ **7.** $x + 2 > 4$

8. $x - 3 < 6$ **9.** $x - 6 \leqslant -2$ **10.** $x + 4 \geqslant -1$ **11.** $3x < 2x - 4$ **12.** $4x > 3x + 2$

13. $4 < x$ **14.** $-2 \leqslant x$ **15.** $-6 > x + 1$ **16.** $2 < x - 3$ **17.** $6 \leqslant x + 4$

18. $-2 \leqslant x - 5$

Use set notation, $\{x \mid \_\_\_\_\}$, to represent the set of numbers with the given graphs.

19. ... −5 0 5

20. ... −5 0 5

Answers

4.

5. 

6.

7. 

8.

9. 

10.

11.

12.

13.

14.

15.

16.

17.

18.

19. $\{x \mid x > -3\}$ 20. $\{x \mid x < 2\}$ 21. $\{x \mid x < -7\}$ 22. $\{x \mid x > 21\}$

23. $\{x \mid x > 34\}$ 24. $\{x \mid x < -16\}$

SELF-EVALUATION TEST, FORM A

1. The solution of $\frac{x}{4} = -2$ is______.

2. A solution of an equation is sometimes called a_________of the equation.

3. When $d = rt$ is solved for t, the result is ________.

4. The solution of $3y + 4 = y - 6$ is______.

5. Find three consecutive integers whose sum is -18.
   The integers are_____ ,______ , and_______.

6. To transform $\frac{x}{3} = 6$ to $x = 18$, the_______________ axiom for equations is applied.

7. The solution of $\frac{2}{3}x + 2 = -6$ is_______.

8. If all the solutions of one equation are solutions of another, and vice versa, the equations are said to be_____________.

9. When $x + 3y = c$ is solved for y, the result is____________.

10. The solution of $y - 4 = 7 - 2y + 22$ is______.

11. Find the dimensions of a rectangle that is six times as long as it is wide and whose perimeter is 294 feet. The rectangle is______feet wide and________feet long.

12. When $\frac{by}{c} - a = 0$ is solved for y, the result is_______.

13. Let x represent an element of the set of integers, and graph the solution set of $x \geq 9$ on the line provided.

    −15 −10 −5 0 5 10 15

14. Let x represent an element of the set of integers, and graph the solution set of $x - 3 < 2x$ on the line provided.

    −5 0 5

15. If x is an integer, graph $\{x \mid x > -1\}$ on the line provided.

    −5 0 5

SELF-EVALUATION TEST, FORM B

1. The solution of $5x - 1 = x + 11$ is_____.

2. If all the solutions of one equation are solutions of another, and vice versa, the equations are said to be_______________.

3. The solution of $\frac{3}{4}x - 2 = -8$ is______.

4. Find three consecutive integers whose sum is -39. The integers are_______, _______, and________.

5. When $\frac{ax}{c} + b = 0$ is solved for x the result is_________.

6. To transform $\frac{y}{2} = 10$ to $y = 20$, the_________________axiom for equations is applied.

7. When $d = rt$ is solved for r, the result is________.

8. A solution of an equation is sometimes called a________of the equation.

9. The solution of $\frac{y}{3} = -5$ is________.

10. When $x + 2y = b$ is solved for y, the result is____________.

11. The solution of $2x + 3 = 2 - 3x + 16$ is_____.

12. There were 68,250 votes cast in a recent election. If the winning candidate received 320 more votes than the losing candidate, how many votes did the winning candidate receive?

13. Let x represent an element of the set of integers, and graph the solution of $x \leq 5$ on the line provided.

−15 −10 −5 0 5 10 15

14. Let x represent an element of the set of integers, and graph the solution set of $x + 2 < 2x$ on the line provided.

−5 0 5

15. If x is an integer, graph $\{x \mid x > -3\}$ on the line provided.

−5 0 5

ANSWERS TO TESTS

Prerequisite Test

1. 5
2. 13
3. 16
4. negative
5. -4
6. -9
7. negative
8. 16
9. 21
10. 5

Form A

1. -8
2. root
3. $t = \frac{d}{r}$
4. -5
5. $-7, -6, -5$
6. multiplication
7. -12
8. equivalent
9. $y = \frac{c - x}{3}$
10. 11
11. 21, 126
12. $y = \frac{ac}{b}$
13. $-15 \quad -10 \quad -5 \quad 0 \quad 5 \quad 10 \quad 15$
14. $-5 \quad 0 \quad 5$
15. $-5 \quad 0 \quad 5$

Form B

1. 3
2. equivalent
3. -8
4. $-14, -13, -12$
5. $x = \frac{-bc}{a}$
6. multiplication
7. $r = \frac{d}{t}$
8. root
9. -15
10. $y = \frac{b - x}{2}$
11. 3
12. 34,285 votes
13. 

14. $-5 \quad 0 \quad 5$
15. 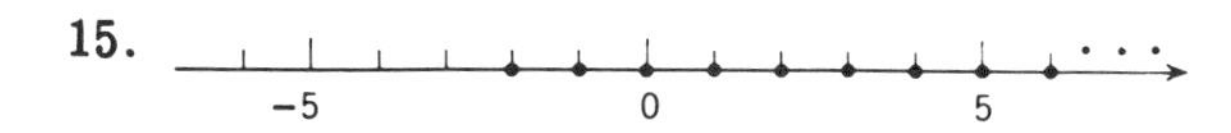

VOCABULARY, UNIT III

The frame in which each word is introduced is shown in parentheses.

addition axiom (70)

addition axiom for inequalities (443)

coefficient (111)

consecutive integers (285)

division axiom (104)

equations (1)

equivalent equations (27)

even integers (295)

first-degree equations (34)

formula (324)

greater than (390)

inequality (401)

integers (281)

left member (3)

less than (386)

line graph (394)

multiplication axiom (163)

odd integers (301)

right member (3)

root (9)

set notation (464)

solution (7)

symmetric law of equality (213)

UNIT IV Products and Factors

OBJECTIVES

Upon completion of the unit, the student should:

1. Be able to apply the distributive law to express the products of combinations of monomials and binomials as polynomials.

2. Be able to factor polynomials (through trinomials) with integral coefficients into products of polynomials with integral coefficients.

3. Be able to solve simple first-degree equations involving parentheses.

4. Be able to solve simple word problems expressible in terms of first-degree equations involving parentheses.

CONTENTS

Unit IV

Products and Factors

Remark. This unit is concerned with factors and products. You should recall that a product is the result of multiplying numbers together. The numbers multiplied together are called the factors of the product. However, the factors and products you are going to be interested in now are somewhat more involved than those you have worked with previously, and are going to center around the application of the distributive law, which is the subject of the first few frames.

16

1. Recall that, in arithmetic, the product $2(3 + 5)$ can be computed by first adding 3 and 5 to obtain 8, and then $2(3 + 5)$ can be written $2(8)$, which is equal to the number______.

16

2. $2(3 + 5)$ can also be computed by multiplying 2 times 3 and 2 times 5 and adding the resulting products. Thus, $2(3) + 2(5) = 6 + 10 =$______.

7; 3

3. The expressions $2(3 + 5)$ and $2(3) + 2(5)$ are equal. Similarly, $5(7 + 3) = 5(\_\_) + 5(\_\_)$.

3; 3

4. $3(6 + 2) = \_\_\_(6) + \_\_\_(2)$.

distributive

5. The equation $5(15 + 7) = 5(15) + 5(7)$ is an example of an application of a general principle of numbers. The principle is called the distributive law, and is stated symbolically

If a, b, and c are integers, $a(b + c) = ab + ac$

Writing $6(10 + 3)$ as $6(10) + 6(3)$ is an application of of the____________________ law.

$5 + 7$

6. The name "distributive" arises from the fact that the equation $a(b + c) = ab + ac$ serves to "distribute" the multiplication by a over the sum $b + c$. In particular, $3(5 + 7)$ "distributes" the multiplication by 3 over the sum____________.

distributive

7. The____________________ law guarantees that $4(5 + 6) = 20 + 24$.

12

You multiply 6 times 2.

8. $6(3 + 2) = 18 +$____.

21

9. $7(5 + 3) = 35 +$____.

60

10. $10(6 - 4) =$____$- 40$.

20

11. $4(8 - 5) = 32 -$____.

$5x + 10$

12. $5(x + 2) =$____________.

$3x - 21$

13. $3(x - 7) =$____________.

4

You multiply 4 times 1.

14. The distributive law can be extended to the form $a(b + c + d + \ldots) = ab + ac + ad + \ldots$.
Thus, $4(x^2 + x + 1) = 4x^2 + 4x +$ ____.

$-21x$

You multiply 7 by the middle term, $-3x$.

15. $7(2x^2 - 3x - 1) = 14x^2$ ______ $- 7$.

$15x^2$

3 times $5x^2$.

16. $3(5x^2 + 2x - 3) =$ ______ $+ 6x - 9$.

$6y + 9x + 12$

17. $3(2y + 3x + 4) =$ ____________.

$3x^2 - 2xy + xz$

18. $x(3x - 2y + z) =$ ____________.

$3a + 6$

19. Another way to look at the process of using the distributive law is to view it as a means of writing a product of two polynomials as another polynomial. Thus, applying the distributive law, the product $3(a + 2)$ can be written as the binomial ________.

$2x^2$; x

20. The product $-x(x^2 - 2x + 1)$ can be written as the trinomial $-x^3 +$ _____ $-$ _____.

$2ax + 2a^2$

21. Write $2a(x + a)$ as a binomial.

$xy^2 - x^2y + xy$

22. Write $xy(y - x + 1)$ as a trinomial.

$6y^3 + 9xy^2 - 3y^2$

Did you think, "$-3y^2$ times $-2y$, $-3y^2$ times $-3x$, and $-3y^2$ times 1"?

23. To "multiply" the factors of $-x(x - y + 1)$ means to write the trinomial $-x^2 + xy - x$. Multiply: $-3y^2(-2y - 3x + 1)$.

distributive

24. Writing the product $x^3(x^2 + x + 3)$ as the polynomial $x^5 + x^4 + 3x^3$ is an application of the ____________ law.

$-2x^3 + 2x^2 - 6x$

25. Apply the distributive law to write $-2x(x^2 - x + 3)$ as a polynomial.

$4y^3 - 8y^2 + 12y$

26. Multiply: $4y(y^2 - 2y + 3)$.

$15xy - 5x^2y + 5xy^3$

27. Multiply: $5xy(3 - x + y^2)$.

$-3x^2 + 2x^3 - x^4$

28. Multiply: $-x^2(3 - 2x + x^2)$.

Remark. The work thus far has been for the purpose of familiarizing you with the distributive law. As you will see, the distributive law is used to help you rewrite expressions in which parentheses occur, as equal expressions that are free of parentheses.

$-a$

29. If the distributive law is applied to $-a(x + 1)$, the expression $-a(x + 1) + ax$ can be written as the polynomial $-ax - a + ax$. This last expression can be simplified by adding like terms to give________.

distributive; adding

Or combining.

30. The expression

$$2a(x + 3) - 3a(x - 3)$$

can be written as the polynomial $2ax + 6a - 3ax + 9a$ by applying the ____________ law, and then simplified, to $-ax + 15a$ by__________ like terms.

$x + 5y$

31. The result of applying the distributive law to $3(x + y) - 2(x - y)$ is $3x + 3y - 2x + 2y$. This expression can be simplified to ______________ by adding like terms.

$x^2 + 4x + 1$

After first applying the distributive law, you have $3x^2 + 6x - 3 - 2x^2 - 2x + 4$.

32. To "simplify"

$$3(x^2 + 2x - 1) - 2(x^2 + x - 2)$$

means to apply the distributive law and then add any like terms in the resulting expression. The result of doing this is ______________.

$-x^2 - 12x + 8$

After first applying the distributive law, you have $2x^2 - 6x + 2 - 3x^2 - 6x + 6$.

33. Simplify:

$$2(x^2 - 3x + 1) - 3(x^2 + 2x - 2).$$

$2y$

34. Simplify:

$$2(x + 3y) - 2x(1 + y) + 2y(x - 2).$$

$(a + b)$

35. Recall that $-a$ is equal to $(-1)(a)$. Thus, $-(3 + x) = -1(3 + x)$, $-(y + 3) = -1(y + 3)$, and $-(a + b) = -1(\_\_\_\_\_\_\_\_)$.

-1

36. $-(x^2 + 2x + 3) =$ ______ $(x^2 + 2x + 3)$.

-1

37. $-(a + 2b - c) =$ ______ $(a + 2b - c)$.

distributive

38. An expression such as $-(2x - y)$ can be written $-1(2x - y)$, and then $-2x + y$ by applying the ______________ law.

distributive; $-2x + y$

39. After writing an expression such as $-(2x - y)$ as $-1(2x - y)$, the ____________________ law can be applied to write _______.

$-a - c$

First write $-1(a + c)$ if it helps you apply the distributive law.

40. Write an expression without parentheses equal to $-(a + c)$.

$-2 + x$

$-2 + x$ is just the negative of $2 - x$.

41. Write an expression without parentheses equal to $-(2 - x)$.

$-2r + s + t$

42. Write an expression without parentheses equal to $-(2r - s - t)$.

$-3x - 2y + z$

43. Write an expression without parentheses equal to $-(3x + 2y - z)$.

$a - 2b + c$

$1a - 2b + 1c$ is also correct—but it is not necessary to show the coefficient "1."

44. If $(a - 2b + c)$ is written with the understood coefficient 1, as $1(a - 2b + c)$, applying the distributive law leads to the sum ________________.

$3a - b + c$

45. Write an expression without parentheses equal to $(3a - b + c)$.

$x + 2y - 3z$

46. Remove the parentheses from $(x + 2y - 3z)$.

Remark. Writing $-(x + 2y - 3)$ as $-x - 2y + 3$ and writing $(x + 2y - 3)$ as $x + 2y - 3$ should already be familiar procedures. We are looking at them, this time, as an application of the distributive law while the first time we worked with such expressions, in Unit II, we simply viewed $-(x + 2y - 3)$ as the negative of $(x + 2y - 3)$ which is $-x - 2y + 3$, and we viewed $(x + 2y - 3)$ and $x + 2y - 3$ as equivalent expressions, each form to be used as convenient.

$x + 3y + 1$

Upon removing parentheses you have $2x + y - 4 - x + 2y + 5$.

47. Simplify: $(2x + y - 4) - (x - 2y - 5)$.

$2x - 3y + 1$

48. Simplify: $(3x - 2y + 4) - (x + y + 3)$.

$x^2 + 7x - 3$

49. Simplify: $(2x^2 + 3x - 2) - (x^2 - 4x + 1)$.

$2x - 6$

50. Simplify: $(x^2 + 2x - 3) + (x - 4) - (x^2 + x - 1)$.

$-x^2 + 5x - 2$

51. Simplify: $(x^2 + 3x - 2) + 3(x - 1) - (2x^2 + x - 3)$.*

Remark. Heretofore, we have been interested solely in applying the distributive law in the form $a(b + c) = ab + ac$. We are now interested in reversing this procedure and applying the distributive law in the form

$$ab + ac = a(b + c).$$

* See Exercise 1, page 60, for additional practice.

distributive

52. The distributive law also permits writing a polynomial as a product. Since

$$a(b + c) = ab + ac,$$

the equality can be reversed, and

$$ab + ac = a(b + c).$$

The polynomial $2x + 2y$ can be written as the product $2(x + y)$ by applying the________________law.

4

53. The polynomial $4x + 4y$ can be written as the product_____ $(x + y)$.

3

54. The polynomial $3a + 3b$ can be written as the product_____$(a + b)$.

$x + 2$

55. The polynomial $2x + 4$ can be written as the product $2($________$)$.

factors

56. The product $a(b + c)$ contains the factors a and $(b + c)$. The product $c(d + e)$ contains the__________ c and $(d + e)$.

3

57. The factors of $3(x - 2)$ are______and $(x - 2)$.

$x + 5$

Note that x itself, or 5 itself, is not a factor. The sum $(x + 5)$ is a factor.

58. The factors of $2(x + 5)$ are 2 and (__________).

$(x + y)$

59. The factors of $x(x + y)$ are x and____________.

$2x$

60. The factors of $2x(x^2 - 2x + 1)$ are_____and $(x^2 - 2x + 1)$.

factoring

61. Since the result of writing a polynomial as a product is a set of factors, the process is called factoring. Thus, writing $2x + 2$ as $2(x + 1)$ is an example of________________.

factored

Because $x(x + 2) = x^2 + 2x$.

62. Since $3x + 9 = 3(x + 3)$, the polynomial $3x + 9$ is said to be factored to the product $3(x + 3)$. The sum $x^2 + 2x$ can be________________to $x(x + 2)$.

3

63. To "factor $ax + ay$" means to write $a(x + y)$. To "factor $3x + 3y$" means to write_____$(x + y)$.

$7x + 7$

64. The result of factoring a polynomial can be checked by determining whether multiplying the factors gives the original expression. Thus, $7(x + 1)$ is a factored form of $7x + 7$, because multiplying 7 by each term in $x + 1$ yields____________.

Yes

Because 2 times x^2 plus 2 times 1 is $2x^2 + 2$.

65. The factorization $2x^2 + 2 = 2(x^2 + 1)$ can be checked by multiplying the factors 2 and $(x^2 + 1)$. Is $2(x^2 + 1)$ a correct factorization of $2x^2 + 2$?

Yes

66. Check the factorization of $ar + br = r(a + b)$ by multiplying the factors in the right member. Is the factorization correct?

2

Because $2(x + 1) = 2x + 2$.

67. The factored form of $2x + 2 =$_____$(x + 1)$.

$x + 1$

Because $3(x + 1) = 3x + 3$.

68. The factored form of $3x + 3$ is $3($________$)$.

$5(x + 1)$

Because $5(x + 1) = 5x + 5$.

69. The factored form of $5x + 5$ is________.

a

70. By the commutative law of multiplication, $a(b + c) = (b + c)a$. Therefore, the factored form of $ab + ac$ can be written either as $a(b + c)$ or as $(b + c)$____.

3

71. By the commutative law of multiplication, $3(x + 1)$ can be written as $(x + 1)$____.

$(x + 2)x$

72. By the commutative law of multiplication, $x(x + 2)$ can be written as________.

$2(x + 3)$

By preferable, we mean that most people write it this way.

73. When writing the factored form of a polynomial, it is customary to write any monomial factor first. Thus, the form $a(x + y)$ is generally preferred to the form $(x + y)a$. Of the two forms $2(x + 3)$ and $(x + 3)2$, it is preferable to write________.

$8(x + 3)$

74. Of the two forms $8(x + 3)$ and $(x + 3)8$, it is preferable to write________.

Remark. Now that you know what it means to "factor" a polynomial, your next task is to learn how to do it. As you continue through the program, if your response contains the same factors as those shown in the response column, but you have them in a different order; that is, if you write (3)(2) instead of (2)(3), don't worry about it, your response is perfectly correct.

factor

75. In order for a polynomial to be factorable, it is necessary that each term in the polynomial contain a common factor. The terms of the expression $3x^2 + 3$ contain 3 as a common________.

do not

There is no whole number or no variable that is a factor of both terms.

76. The common factors to be sought are those consisting of integers and natural number powers of variables. The terms of $3x + 4$ do not contain a common factor, because they contain no integer or variable in common. The terms of $2x + 5$ (do/do not) contain a common factor.

5

77. The terms of $5x^2 + 5$ contain the common factor_____.

$x^2 + 1$

78. The expression $5x^2 + 5$ is written in factored form as 5(___________).

Yes

5 times x^2 plus 5 times 1.

79. In factoring expressions, it is well to establish the habit of checking. Does $5(x^2 + 1) = 5x^2 + 5$?

2

80. The terms of $2x^2 - 4x + 6$ contain the common factor_____.

$-2x$

You ask yourself, "By what do I have to multiply 2 to get $-4x$?

81. $2x^2 - 4x + 6$ can be factored as $2(x^2$ _______ $+ 3)$.

Yes

82. Does $2(x^2 - 2x + 3) = 2x^2 - 4x + 6$?

3

83. The terms of $3x^2 - 6x + 9$ contain the common factor_____.

$x^2 - 2x + 3$

84. $3x^2 - 6x + 9$ can be factored as 3(________________).

Yes

85. Does $3(x^2 - 2x + 3) = 3x^2 - 6x + 9$?

$3y$

You could have said 3 or y, but why not include both?

86. The terms of $2x^3 - 2x^2 + 2x$ contain the common factor $2x$. The terms of $3y^3 - 3y^2 + 3y$ contain the common factor______.

$y^2 - y + 1$

You can ask yourself, "By what do I have to multiply $3y$ to get $3y^3$? $-3y^2$? $3y$?"

87. $3y^3 - 3y^2 + 3y$ can be factored as $3y($______________$)$.

Yes

88. Does $3y(y^2 - y + 1) = 3y^3 - 3y^2 + 3y$?

$(x^2 + 2x + 3)$

89. The terms that belong inside the parentheses in the factored form of

$$2x^2 + 4x + 6 = 2(\qquad\qquad),$$

can be found by asking "By what must I multiply 2 to obtain $2x^2$? $4x$? 6?" The correct factors in this example are 2 and (______________).

Yes

90. Does $2(x^2 + 2x + 3) = 2x^2 + 4x + 6$?

$3x(x^2 - 3x + 4)$

91. The terms that belong inside the parentheses in the factored form of

$$3x^3 - 9x^2 + 12x = 3x(\qquad\qquad),$$

can be found by inquiring "By what must I multiply $3x$ to obtain $3x^3$? $-9x^2$? $12x$? The correct factored form of $3x^3 - 9x^2 + 12x$ is $3x($______________$)$.

Yes

92. Does $3x(x^2 - 3x + 4) = 3x^3 - 9x^2 + 12x$?

$3a$

93. The terms of $3ax + 3ay + 3az$ contain the common factor______.

$x + y + z$

94. $3ax + 3ay + 3az$ can be factored as $3a($______________$)$.

Yes

95. Does $3a(x + y + z) = 3ax + 3ay + 3az$?

$5b(x + y + z)$

96. Factor $5bx + 5by + 5bz$.

$5b$ is a factor of each term of the trinomial.

Yes

97. Does $5b(x + y + z) = 5bx + 5by + 5bz$?

$x - 2$

98. The expression $4x^2 - 8x$ can be factored in a number of ways. For example, as $4(x^2 - 2x)$, as $2(2x^2 - 4x)$, as $x(4x - 8)$, or as $4x($__________$)$.

completely

99. A factored expression of the form $a(x + y)$ is said to be completely factored if the terms within the parentheses contain no common monomial factor. Thus, the expression $4x(x - 2)$ is________________ factored.

is not

100. Since the terms within the parentheses still contain the common factor 2, $3(2x + 4)$ (is/is not) completely factored.

x

101. $2y(x^2 + 2x)$ is not completely factored because the terms within the parentheses still contain the common factor____.

$x + 2$

102. The completely factored form of $2y(x^2 + 2x)$ is $2xy($__________$)$.

Yes

103. Does $2xy(x + 2) = 2x^2y + 4xy$?

$2x(2x + 1)$

104. Factor $4x^2 + 2x$ completely.

Yes

105. Does $4x(2x + 1) = 8x^2 + 4x$?

$3x(x^2 - 2x + 3)$

Did you check by multiplying the factors?

106. Factor $3x^3 - 6x^2 + 9x$ completely.

$2x(2x - 1)$

107. From this point on, the word "factor" will mean "factor completely." Factor $4x^2 - 2x$.

$5x(x^2 + 2x + 3)$

108. Factor $5x^3 + 10x^2 + 15x$.

$6y(y^2 + 2y + 3)$

109. Factor $6y^3 + 12y^2 + 18y$.

y

110. $-3x^2 - 3xy$ can be factored as $-3x(x +$____$)$.

Yes

111. Does $-3x(x + y) = -3x^2 - 3xy$?

$(y - 2)$

112. $-6y^3 + 12y^2$ can be factored as $-6y^2($__________$)$.

$-1(a + b + c)$

113. Factor -1 from $-a - b - c$.

Yes

114. Does $-1(a + b + c) = -a - b - c$?

$x^2 - 2x + 3$

115. $-2x^2 + 4x - 6 = -2(\_\_\_\_\_\_\_\_\_\_)$.

$-x^2 + 2x - 3$

116. $-2x^2 + 4x - 6 = 2(\_\_\_\_\_\_\_\_\_\_)$.

$-x^2 + 2x + 3$

117. $-2x^2 + 4x - 6$ can be factored as $-2(x^2 - 2x + 3)$ or $2(-x^2 + 2x - 3)$. That is, either -2 or 2 can be factored from each term. The expression $-3x^2 + 6x + 9$ can be factored as $-3(x^2 - 2x - 3)$ or $3(\_\_\_\_\_\_\_\_\_\_)$.

Remark. Since in factoring a monomial from a polynomial we have a choice of signs for the monomial, how shall the choice be made? That is, should $-2x^2 + 4x - 6$ be factored as $-2(x^2 - 2x + 3)$ or as $2(-x^2 + 2x - 3)$? The answer is that it does not make any difference. However, to avoid having to make two responses from here on, we shall agree to factor so that the first term in the polynomial factor has a positive sign. Thus, above, $-2(x^2 - 2x + 3)$ is the way in which the response would appear in the response column.

$-x(x + 1)$

118. Factor $-x^2 - x$.

$-3(x^2 - 2x + 3)$

119. Factor $-3x^2 + 6x - 9$.

$-xy(y - 1 + x)$

120. Factor $-xy^2 + xy - x^2y$.

$x(x^2 - 2x + 3)$

121. Factor x from $x^3 - 2x^2 + 3x$.

$3x^3(x^2 + 3x + 4)$

122. Factor $3x^3$ from $3x^5 + 9x^4 + 12x^3$.

$2xy(y + 2x + 3)$

Are you checking?

123. Factor $2xy^2 + 4x^2y + 6xy$.

$-a(bc + b + c)$

124. Factor $-abc - ab - ac$.

$-xy(6x - y + 3)$

125. Factor $-6x^2y + xy^2 - 3xy$.

$12y^2(y^2 - 2y + 3)$

126. Factor $12y^4 - 24y^3 + 36y^2$. *

Remark. Factoring monomials from polynomials is one type of factoring. There are other types of factors for products, however, and these shall be our next concern. As a first consideration, we shall look again at the distributive law, and some results of applying it twice.

binomial

127. Recall that a polynomial containing two terms is called a binomial. $(a + b)$ is a binomial. $(c + d)$ is a__________________.

product

128. The product of two binomials can be indicated by writing the binomials side by side. Thus, $(a + b)(2a - b)$ is the__________________of two binomials.

$2x - y$

Both x and y have to be multiplied by this expression.

129. The distributive law permits $(a + b)(2a - b)$ to be written $a(2a - b) + b(2a - b)$, where each term in the first binomial is multiplied by the second binomial. Similarly, $(x + y)(2x - y) = x(2x - y) + y($__________$)$.

* See Exercise 2, page 60, for additional practice.

$3r - s$; $3r - s$

130. $(r + s)(3r - s) = r(\_\_\_\_\_\_\_\_) + s(\_\_\_\_\_\_\_\_)$.

$x + 2$; $x + 2$

131. $(2x - 1)(x + 2) = 2x(\_\_\_\_\_\_\_\_) - 1(\_\_\_\_\_\_\_\_)$.

$3x$; 2

132. $(3x - 2)(x + 1) = \_\_\_\_(x + 1) - \_\_\_\_(x + 1)$.

$6x$; 3

133. $(x + 3)(2x + 1) = x(2x + 1) + 3(2x + 1)$. Applying the distributive law again, the right member of this equation can be written as the sum of four terms $2x^2 + x + \_\_\_\_\_\_ + \_\_\_\_$.

$7x$

The like terms are x and $6x$.

134. Like terms in the expression $2x^2 + x + 6x + 3$ can be combined to give $2x^2 + \_\_\_\_\_\_ + 3$.

$x^2 + 3x + 2x + 6$

135. $(x + 2)(x + 3) = x(x + 3) + 2(x + 3)$. Applying the distributive law again, the right member of this equation can be written as the sum of four terms ____________.

$x^2 + 5x + 6$

136. Like terms in the expression $x^2 + 3x + 2x + 6$ can be combined to give the expression ________ consisting of three terms.

$ac + ad + bc + bd$

137. Application of the distributive law to the product of two binomial factors of the form $(a + b)(c + d)$ gives $a(c + d) + b(c + d)$. Applying the distributive law again, this latter expression can be written as the sum of four terms ____________.

$x^2 + 4x - 3x - 12$

You first have $x(x + 4) - 3(x + 4)$.

138. Apply the distributive law twice, and represent the product $(x - 3)(x + 4)$ as the sum of four monomial terms.

$x^2 + x - 12$

You just combine like terms.

139. Represent $x^2 + 4x - 3x - 12$ as the sum of three monomial terms.

$x^2 - 2x - x + 2$

You first have $x(x - 2) - 1(x - 2)$.

140. Apply the distributive law twice and represent the product $(x - 1)(x - 2)$ as the sum of four monomial terms.

$x^2 - 3x + 2$

141. Represent $x^2 - 2x - x + 2$ as the sum of three monomial terms.

$3x$; $4x$

142. When multiplying two binomials, the resulting four terms can be written directly by multiplying each term in one binomial by each term in the other. Using the order previously used, the multiplication is performed as follows;

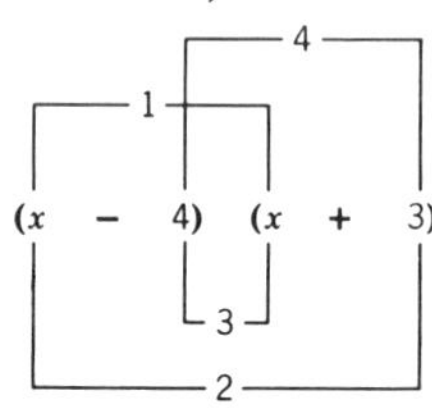

This leads to the four terms
$x^2 +$ ______ $-$ ______ $- 12$.

$x^2 - x - 12$

143. Upon combining like terms, $x^2 + 3x - 4x - 12$ can be written ________________.

$x^2 + 5x + 4x + 20$

144. Represent $(x + 4)(x + 5)$ as the sum of four terms.

$x^2 + 9x + 20$

145. Upon combining like terms, $x^2 + 5x + 4x + 20$ can be written ________________ .

$6x$

You add x and $5x$ mentally to obtain $6x$.

146. With practice, some products of binomials can be written directly as the sum of three terms (that is, as a trinomial) by combining the like terms mentally. Thus, $(x + 5)(x + 1) = x^2 +$ ______ $+ 5$.

$6x$

$3x + 3x$.

147. $(x + 3)(x + 3) = x^2 +$ ______ $+ 9$.

x

$3x - 2x$.

148. $(x + 3)(x - 2) = x^2 +$ ______ $- 6$.

$-3x$

149. $(x + 5)(x - 8) = x^2 -$ ______ $- 40$.

$8x$

150. $(x + 4)(x + 4) = x^2 +$ ______ $+ 16$.

$3x$

151. $(x - 5)(x + 8) = x^2 +$ ______ $- 40$.

$+ 2x$

If you are having trouble, write $x^2 - x + 3x - 3$, then combine like terms.

152. $(x + 3)(x - 1) = x^2$ ______ $- 3$.

$x^2 - 2x - 15$

You first have $x^2 - 5x + 3x - 15$.

153. Write $(x + 3)(x - 5)$ as a trinomial.

$x^2 - 6x - 7$

154. Write $(x - 7)(x + 1)$ as a trinomial.

$x^2 - 8x - 9$

155. Write $(x + 1)(x - 9)$ as a trinomial.

$x^2 + 3x - 28$

156. Write $(x + 7)(x - 4)$ as a trinomial.

$y^2 - 3y - 70$

157. Multiply: $(y + 7)(y - 10)$.

$2x^2 - 5x - 12$

158. Multiply: $(x - 4)(2x + 3)$.

$2x^2 + 7x + 5$

159. Write $(2x + 5)(x + 1)$ as a trinomial.

$x^2 + bx - 6b^2$

You first have $x^2 + 3bx - 2bx - 6b^2$.

160. Write $(x - 2b)(x + 3b)$ as a trinomial.

$x^2 - 3xy + 2y^2$

161. Write $(x - y)(x - 2y)$ as a trinomial.

$x^2 + 4xz + 3z^2$

162. Multiply: $(x + z)(x + 3z)$.

$x^2 - 3xy - 10y^2$

163. $(x + 2y)(x - 5y) =$ ____________________.

$a^2 + 4ab - 5b^2$

164. $(a - b)(a + 5b) =$ ____________________.

Remark. You are now able to apply the distributive law to find the product of two binomials. The process you have learned is applicable to any two binomials. There are two special cases of this process that occur frequently enough in algebra to make it worthwhile looking at them in some detail. The first such case has to do with squaring a binomial.

$x + 3$

165. The power $(x + 3)^2$ can be written as the product of two equal factors $(x + 3)(\_\_\_\_\_\_)$.

$x^2 + 6x + 9$

166. $(x + 3)(x + 3)$ is equal to the trinomial ____________.

$(x - 2)(x - 2)$

167. Write $(x - 2)^2$ as the product of two equal factors.

$x^2 - 4x + 4$

168. Write $(x - 2)(x - 2)$ as a trinomial.

$(x - 5)(x - 5)$

169. $(x - 5)^2 = (\_\_\_\_\_\_)(\_\_\_\_\_\_)$.

$x^2 - 10x + 25$

170. Write $(x - 5)(x - 5)$ as a trinomial.

$(3x - y)(3x - y)$

171. $(3x - y)^2 = (\_\_\_\_\_\_)(\_\_\_\_\_\_)$.

$9x^2 - 6xy + y^2$

172. Write $(3x - y)(3x - y)$ as a trinomial.

$4x^2 + 4x + 1$

You first have $(2x + 1)(2x + 1)$.

173. Write $(2x + 1)^2$ as a trinomial.

$4x^2 - 4x + 1$

174. Write $(2x - 1)^2$ as a trinomial.

$4x^2 - 4x + 1$

175. $(2x - 1)^2$ says, "square $(2x - 1)$." The square of $(2x - 1)$, when written as a trinomial, is ______________________.

$4x^2 + 12x + 9$

176. Square $(2x + 3)$.

$4x^2 - 4xy + y^2$

177. Square $(2x - y)$.

$a^2 - 6ab + 9b^2$

178. Square $(a - 3b)$.

Remark. If we wanted to formalize the procedure used in squaring a binomial such as $(a + b)$, we could say that the square of $a + b$ consists of the square of the first term, a, plus twice the product of the terms a and b, plus the square of the second term, b. That is, $(a + b)^2 = a^2 + 2ab + b^2$. However, since this is nothing but a special case of the more general procedure of multiplying two binomials, special rules are not important.

We shall next look at another special case of multiplying two binomials.

difference

179. Another special case of the product of two binomials is that where the binomials represent the sum and the difference of the same numbers. For example, in the product $(x + 1)(x - 1)$, $x + 1$ represents the sum of x and 1, and $x - 1$ represents the ______________________ of x and 1.

sum; difference

180. In the product $(x + 2y)(x - 2y)$, $x + 2y$ represents the __________ of x and $2y$, and $x - 2y$ represents the ______________________ of x and $2y$.

sum; difference

181. $(x + y)(x - y)$ represents the product of the ________ of x and y and the ________ of x and y.

$x^2 - y^2$

182. The product $(x + y)(x - y)$ can first be written as four terms, $x^2 - xy + xy - y^2$. When like terms are combined, the result is ________.

$x^2 - 4$

183. $(x + 2)(x - 2) = x^2 - 2x + 2x - 4$. When the like terms are combined, the result is ________.

squares

184. The product of the sum and difference of two numbers is always the difference of the squares of the numbers. Thus, $(x - 5)(x + 5)$ is equal to $x^2 - 25$ which is the difference of the ________ of x and 5.

$x^2 - 9$

185. $(x + 3)(x - 3) =$ ________.

The difference of the squares of x and 3.

$x^2 - 49$

186. $(x + 7)(x - 7) =$ ________.

$4x^2 - 1$

187. $(2x + 1)(2x - 1) =$ ________.

$9x^2 - 4y^2$

188. $(3x - 2y)(3x + 2y) =$ ________.

$25a^2 - b^2$

189. $(5a + b)(5a - b) =$ ________.

Are you squaring the first term in a factor, and then subtracting the square of the second term?

$9a^2 - 25c^2$

190. $(3a - 5c)(3a + 5c) =$ ________________ .

$a^2x^2 - b^2$

191. $(ax - b)(ax + b) =$ ________________ .

Remark. Remember that squaring a binomial, or finding the product of the sum and difference of two numbers, are both special cases of a more general process, that of multiplying two binomials. In the following frames you will learn to find the product of a monomial and two binomials.

$3x - 18$

192. If a monomial is to be multiplied by two binomials, for example, $3(x - 2)(x + 3)$, it is best to multiply the binomials first to obtain $3(x^2 + x - 6)$. This product can then be written as the trinomial $3x^2 +$ ____________ .

$2x^2 + 6x + 4$

You would first have $2(x^2 + 3x + 2)$.

193. Multiply: $2(x + 1)(x + 2)$.

$a^3 + 4a^2 - 5a$

The first step is $a(a^2 + 4a - 5)$.

194. Write $a(a - 1)(a + 5)$ as a trinomial.

$6x^2 - 16x - 6$

195. Multiply: $2(3x + 1)(x - 3)$.

$6y^2 + 33y + 15$

196. Multiply: $3(2y + 1)(y + 5)$.

$2ax^2 - ax - 6a$

197. $a(x - 2)(2x + 3) =$ ________________ .

$6x^3 - 32x^2 + 10x$

198. $2x(x - 5)(3x - 1) =$ ____________________.

You would first have $2x(3x^2 - 16x + 5)$.

$5x^2 + 20x + 20$

199. $5(x + 2)^2 =$ ____________________.

You first have $5(x^2 + 4x + 4)$.

$6x^2 - 36x + 54$

200. $6(x - 3)^2 =$ ____________________.

$4x^3 - 12x^2 + 9x$

201. $x(2x - 3)^2 =$ ____________________.

$3ax^2 - 24ax + 48a$

202. $3a(x - 4)^2 =$ ____________________.

$x^3 - xy^2$

203. $x(x - y)(x + y) =$ ____________.

$2x^2 - 8$

204. $2(x + 2)(x - 2) =$ ____________. *

Remark. We are now going to look at the reverse of the process we have been studying. We have been using the distributive law to multiply binomials, and binomials and monomials, but now we are going to look at the process of factoring trinomials into binomial factors. Thus, you are going to learn to write trinomials such as $x^2 - 7x + 10$ in the form $(x - 5)(x - 2)$.

$x + 2$

205. Since $(x + 2)(x + 1) = x^2 + 3x + 2$, this equality can be reversed and written $x^2 + 3x + 2 = (x + 2)(x + 1)$. By the commutative law, the right member could also be written $(x + 1)($____________$)$.

* See Exercise 3, page 61, for additional practice.

$(x - 2)(x + 3)$

$(x + 3)(x - 2)$ is also correct; the order doesn't matter.

206. Since $(x - 2)(x + 3) = x^2 + x - 6$, you can reverse the equality and obtain
$x^2 + x - 6 =$ ____________________.

$x - 2$

207. The process of writing a trinomial as the product of two binomials, for example, $x^2 + x - 6$ as $(x + 3)(x - 2)$, is called "factoring the trinomial." The factors of $x^2 + x - 6$ are $x + 3$ and ____________.

product

208. Factoring a trinomial is the process of writing a sum as a product. The sum $x^2 + 3x + 2$ can be written as the ____________________ $(x + 1)(x + 2)$.

$x^2 + 4x + 3$

209. The factors of the trinomial $x^2 + 4x + 3$ are $(x + 3)$ and $(x + 1)$ because
$(x + 3)(x + 1) =$ ____________________.

$(x - 3)(x + 1)$

210. The factors of $x^2 - 2x - 3$ are $(x - 3)$ and $(x + 1)$ because (__________)(__________) $= x^2 - 2x - 3$.

factored

211. $x^2 + 7x + 12$ is said to be factored if we can find two binomials whose product is $x^2 + 7x + 12$. When $x^2 + 7x + 12$ is written as $(x + 4)(x + 3)$ the trinomial has been ____________________.

x

212. The factored form of $x^2 + 7x + 12$ is $(x + 3)($____$+ 4)$, where the product of the first terms of each binomial equals x^2, the first term of the trinomial.

x

213. $x^2 - 3x - 10 = ($____$- 5)(x + 2)$.

x; x

214. $x^2 - 8x + 16 = (\_\_\_ - 4)(\_\_\_ - 4)$.

1

215. The factored form of $x^2 + 6x + 5$ is $(x + 5)(x + \_\_\_)$, where the product of the last term of each binomial equals 5, the last term of the trinomial.

1; 11

216. $x^2 + 12x + 11 = (x + \_\_\_)(x + \_\_\_)$.

Or 11; 1. We will not show the alternative responses in the following frames.

1; 7

217. $x^2 - 8x + 7 = (x - \_\_\_)(x - \_\_\_)$.

x; 2

218. If the trinomial $x^2 + 3x + 2$ is written as the product of two binomials in the form (   )(   ), the first term in each binomial would be_____ and the last terms would be 1 and_____.

x; 1; 1

219. If the trinomial $x^2 + 2x + 1$ is written as the product of two binomials in the form (   )(   ), the first term in each binomial would be______ and the last terms would be ____ and____.

1; 1

220. The trinomial $x^2 + 2x + 1$ is equal to the product $(x + \_\_\_)(x + \_\_\_)$.

You have to have two numbers whose product is 1.

1; 3

221. The trinomial $x^2 + 4x + 3$ is equal to the product $(x + \_\_\_)(x + \_\_\_)$.

1; 5

222. $x^2 + 6x + 5 = (x + \_\_\_)(x + \_\_\_)$.

6

223. In factoring certain trinomials, for example, $x^2 + 6x + 5 = (x + 5)(x + 1)$, the sum of the last terms in the factors is the coefficient of the middle term of the trinomial. In this example, the coefficient of the middle term is_____.

3; 1

224. The factors of $x^2 + 4x + 3$ are $(x + 3)(x + 1)$, and the coefficient of the middle term, 4, is the sum of_____ and_____ , while the last term, 3, is the product of the same numbers.

5; 4

That is, $4 + 1 = 5$, and $4 \cdot 1 = 4$.

225. The trinomial $x^2 + 5x + 4$ has as factors $(x + 4)(x + 1)$, where the sum of the last terms in the factors is_____ and their product is_____.

negative

226. The product of two numbers having the same sign is positive, while the product of two numbers of opposite sign is_______________.

like

They have to be alike, if their product is to be positive.

227. Since the last term in the trinomial $x^2 + 5x + 6$ is positive, the last terms of the factors of this trinomial must have (like/opposite) signs.

like

228. Since the last term in the trinomial $x^2 - 5x + 6$ is positive, the last terms of the factors of this trinomial must have (like/opposite) signs.

opposite

They have to be opposite, if their product is to be -15.

229. Since the last term in the trinomial $x^2 - 2x - 15$ is negative, the last terms of the factors of this trinomial must have (like/opposite) signs.

+; +

230. The factors of $x^2 + 3x + 2$ are $(x$ ___ $1)(x$ ___ $2)$, as can be verified by noting that the product of the last two terms is 2 while their sum is 3.

−; −

231. The factors of $x^2 - 3x + 2$ are $(x$ ___ $1)(x$ ___ $2)$, as can be verified by noting that the product of the last two terms is 2 while their sum is -3.

−

232. The factors of $x^2 - 3x - 4$ are $(x + 1)(x$ ___ $4)$, as can be verified by noting that the product of the last two terms is -4 while their sum is -3.

+; +

233. The factors of $x^2 + 4x + 3$ are $(x$ ___ $1)(x$ ___ $3)$.

Yes

This is the check.

234. Does $(x + 1)(x + 3) = x^2 + 4x + 3$?

−; −

235. The factors of $x^2 - 6x + 8$ are $(x$ ___ $2)(x$ ___ $4)$.

Yes

236. Does $(x - 2)(x - 4) = x^2 - 6x + 8$?

−

The signs have to be opposite if the last term is -10.

237. $x^2 - 3x - 10 = (x + 2)(x$ ___ $5)$.

Yes

238. Does $(x + 2)(x - 5) = x^2 - 3x - 10$?

+; −

Because $(+3)(-5) = -15$ and $+3x - 5x = -2x$.

239. $x^2 - 2x - 15 = (x \_\_\_ 3)(x \_\_\_ 5)$.

Yes

240. Does $(x + 3)(x - 5) = x^2 - 2x - 15$?

+1; +7

Because $(+1)(+7) = 7$ and $+x + 7x = 8x$.

241. $x^2 + 8x + 7 = (x \_\_\_\_\_\_)(x \_\_\_\_\_\_)$.

$(x + 1)(x + 5)$

Check it.

242. $x^2 + 6x + 5 = (\_\_\_\_\_\_)(\_\_\_\_\_\_)$.

3

Because $2 \cdot 3$ also equals 6.

243. It may be necessary to consider more than one set of possibilities for last terms for the factors of a trinomial. As factors for $x^2 + 5x + 6$, it is necessary to consider both $(x - 1)(x + 6)$ and $(x + 2)(x + \_\_\_)$, even though only one set of factors is correct.

2; 3

244. The sum of the last terms of the factors of $x^2 + 5x + 6$ must equal 5, and must have like signs. Thus, $(x + \_\_\_\_)(x + \_\_\_\_)$ is the correct factored form.

$(x - 2)(x - 3)$

$(x - 1)(x - 6)$ won't do because $-x - 6x$ is not equal to $-5x$.

245. The factors of $x^2 - 5x + 6$ must be either $(x - 2)(x - 3)$ or $(x - 1)(x - 6)$ because the last term of the trinomial is +6 and the middle term is negative. Since the middle term is −5, the correct factors are $(\_\_\_\_\_\_)(\_\_\_\_\_\_)$.

$(x - 1)(x - 6)$

246. The factored form of $x^2 - 7x + 6$ must be either $(x - 2)(x - 3)$ or $(x - 1)(x - 6)$ because the last term of the trinomial is positive. The correct factored form is $(\qquad)(\qquad)$.

$(x + 2)(x + 5)$

247. The factored form of $x^2 + 7x + 10$ must be either $(x + 2)(x + 5)$ or $(x + 1)(x + 10)$. The correct factored form is $(\qquad)(\qquad)$.

$(x + 2)(x - 5)$

248. The factored form of $x^2 - 3x - 10$ must be either $(x + 2)(x - 5)$ or $(x + 1)(x - 10)$. The correct factored form is $(\qquad)(\qquad)$.

$(x + 5)(x - 7)$

249. Factor $x^2 - 2x - 35$.

$(x + 5)(x + 7)$

250. Factor $x^2 + 12x + 35$.

$(x - 2)(x - 6)$

251. Factor $x^2 - 8x + 12$.

$(x + 8)(x - 9)$

252. Factor $x^2 - x - 72$.

$(x - 9)(x + 5)$

253. Factor $x^2 - 4x - 45$.

$(x + 3)(x - 15)$

254. $x^2 - 12x - 45 = (\qquad)(\qquad)$.

Remark. The trinomials we have considered thus far have all been factorable. As you will see in the next few frames, however, not all trinomials have binomial factors.

Yes

Its factors are $(x-1)(x-2)$.

255. $x^2 + 10x + 1$ will not factor because there exist no whole numbers whose sum is 10 and whose product is 1. Will $x^2 - 3x + 2$ factor?

is not

Not this one.

256. $x^2 - 3x - 7$ (is/is not) factorable.

No

257. Is $x^2 - x - 11$ factorable?

Yes

Its factors are $(x + 8)(x - 9)$.

258. Is $x^2 - x - 72$ factorable?

Remark. You will remember that when we were studying the multiplication of two binomials, we considered two special cases, the square of a binomial and the product of the sum and the difference of two numbers. We shall next deal with these special cases as polynomials to be factored.

$x^2 - 4$

Remember, the difference of the square of x and the square of 2.

259. If the last terms of two binomials differ only in sign, the product of the binomials will have only two terms. Thus, $(x - 2)(x + 2)$ have last terms that differ only in sign, and their product is __________.

$x^2 - 1$

260. $(x - 1)(x + 1) =$ __________.

$(x + 3)$; $(x - 3)$

261. Because $(x + 3)(x - 3) = x^2 - 9$, the factors of $x^2 - 9$ are (__________) and (__________).

$x^2 - a^2$

262. $(x - a)(x + a) =$ __________.

$(x - 4)$

263. The difference of two squares, $x^2 - a^2$, will always factor into two factors of the form $(x + a)(x - a)$. The factors of $x^2 - 16$ are $(x + 4)$ and (__________).

$(x + 5)(x - 5)$

264. Factor $x^2 - 25$.

$(x + 9)(x - 9)$

265. Factor $x^2 - 81$.

Are you checking?

3; 3

266. The factors of $x^2y^2 - 9$ are $(xy +$ ____) and $(xy -$ ____).

$(xy + 4)(xy - 4)$

267. Factor $x^2y^2 - 16$.

$(x + 2y)$

268. The factors of $x^2 - 4y^2$ are (__________) and $(x - 2y)$.

$(x + 3y)(x - 3y)$

269. Factor $x^2 - 9y^2$.

$x + 1$

270. Since $(x + 1)(x + 1) = x^2 + 2x + 1$,
$x^2 + 2x + 1 = (x + 1)(x + 1) =$ (__________)^2.

$x + 3$

271. A trinomial of the form $x^2 + 2ax + a^2$ is always the square of a binomial, $(x + a)^2$.
$x^2 + 6x + 9 =$ (__________)^2.

$x + 4$

272. $x^2 + 8x + 16 = (\_\_\_\_\_\_\_\_)^2$.

$x - 1$

273. $x^2 - 2x + 1 = (\_\_\_\_\_\_\_\_)^2$.

$(x - 5)(x - 5)$

Or $(x - 5)^2$.

274. Factor $x^2 - 10x + 25$.

$(x - 6)^2$

275. Write $x^2 - 12x + 36$ as the square of a binomial.

Remark. Sometimes the terms in a trinomial have to be rearranged before the factors can be determined. The next sequence of frames will make clear what this means, and explain how to rearrange the terms in a trinomial so that its factors can be more readily identified.

-8; $+x^2$

276. Trinomials may be written in any arrangement of terms. Thus, $x^2 + 2x - 8$ may be written $2x + x^2$_____ , $2x - 8$_____, or any other possible arrangement.

standard

277. If the terms of a trinomial are arranged in descending powers of a variable, it is said to be in standard form. $x^2 + 2x - 8$ is in________________ form.

$x^2 + 3x - 2$

Notice the descending powers of x.

278. The trinomial $3x - 2 + x^2$, when written in standard form, would appear as________________.

$x^2 - 2x + 3$

279. Write $3 - 2x + x^2$ in standard form.

$x^2 + 3x - 10$

280. A trinomial should be written in standard form before attempting to find binomial factors. Thus, before factoring $3x + x^2 - 10$, it should first be written ______________ .

$(x - 2)(x + 5)$

281. Factor $x^2 + 3x - 10$.

$(x + 1)(x - 5)$

282. In factoring $-4x + x^2 - 5$, the trinomial should first be rewritten $x^2 - 4x - 5$, and then (__________)(__________).

$(x + 2)(x - 6)$

You would first rewrite as $x^2 - 4x - 12$.

283. Factor $-4x + x^2 - 12$. (First rewrite the trinomial in standard form.)

$(y - 4)(y + 5)$

Did you first rewrite as $y^2 + y - 20$?

284. Factor $y - 20 + y^2$.

$(y - 3)(y + 8)$

285. Factor $-24 + 5y + y^2$.

$x^2 - 9$

286. Binomials are also said to be in standard form if the terms are arranged in descending powers of the variable. In standard form, $-9 + x^2$ would be written__________.

$(x - 3)(x + 3)$

The difference of two squares, $x^2 - 9$, remember?

287. Factor $x^2 - 9$.

$x^2 - 16$

288. Write $-16 + x^2$ in standard form.

$(x - 4)(x + 4)$

289. Factor $x^2 - 16$.

$(x - 5)(x + 5)$

In standard form the binomial appears as $x^2 - 25$.

290. Factor $-25 + x^2$. (First rewrite the binomial in standard form.)

$(y - 6)(y + 6)$

291. Factor $-36 + y^2$.*

Remark. All of the binomials and trinomials with which we have thus far worked have had leading coefficients of 1. That is, we have worked with trinomials such as $x^2 + 2x - 3$ or with binomials such as $x^2 - 4$. We wish now to consider binomials and trinomials whose leading coefficients are different from 1.

distributive

292. To multiply the binomials $(2x + 3)(x + 1)$, the distributive law is applied to yield $2x(x + 1) + 3(x + 1)$, which is written $2x^2 + 2x + 3x + 3$ by another application of the ____________ law.

$2x^2 + 5x + 3$

By combining like terms.

293. $2x^2 + 2x + 3x + 3$ can be simplified to ________.

$(2x + 3)(x + 1)$

294. To factor an expression of the form $2x^2 + 5x + 3$, it is necessary to consider more than one possibility. Thus, $(2x + 1)(x + 3)$ and $(2x + 3)(x + 1)$ must both be considered. The middle term of the trinomial, $5x$, then determines that (________)(________) is the correct factorization.

* See Exercise 4, page 62, for additional practice.

Yes

295. Does $(2x + 3)(x + 1) = 2x^2 + 5x + 3$?

We are checking again here.

$(2x + 1)(x + 3)$

296. In factoring $2x^2 + 7x + 3$, the pairs of factors $(2x + 1)(x + 3)$ and $(2x + 3)(x + 1)$ must both be considered. The middle term of the trinomial, $7x$, then determines that (______)(______) is the correct factorization.

Yes

297. Does $(2x + 1)(x + 3) = 2x^2 + 7x + 3$?

1; 2

298. $3x^2 + 7x + 2 = (3x +$ ____ $)(x +$ ____ $)$.

Yes

299. Does $(3x + 1)(x + 2) = 3x^2 + 7x + 2$?

2; 1

300. $3x^2 + 5x + 2 = (3x +$ ____ $)(x +$ ____ $)$.

$(2x + 5)(x + 1)$

301. $2x^2 + 7x + 5 =$ (______)(______).

It might be well to remember that $(x + 1)(2x + 5)$ is just as good, the order is not important.

$(2x + 1)(x + 5)$

302. Factor $2x^2 + 11x + 5$.

$(5x + 2)(x + 1)$

303. $5x^2 + 7x + 2 =$ (______)(______).

$(5x + 1)(x + 2)$

304. $5x^2 + 11x + 2 =$ (______)(______).

$-\,;\,-$

305. If the middle term of a trinomial is negative while the signs on the other terms are positive, both binomial factors must contain negative signs. $3x^2 - 4x + 1 = (3x\_\_\_\ 1)(x\_\_\_1)$.

$(2x - 5)(x - 1)$

306. $2x^2 - 7x + 5 = (\_\_\_\_\_\_\_\_)(\_\_\_\_\_\_\_\_)$.

$(5x - 1)(x - 3)$

307. $5x^2 - 16x + 3 = (\_\_\_\_\_\_\_\_)(\_\_\_\_\_\_\_\_)$.

$(5x - 3)(x - 1)$

308. Factor $5x^2 - 8x + 3$.

positive

309. Each of the binomial factors of $5x^2 - 8x + 3$, namely, $(5x - 3)$ and $(x - 1)$, contain minus signs, because the sign on the last term of the trinomial is __________ and the coefficient of the middle term is negative.

negative

310. The binomial factors of $3x^2 + 5x - 2$, $(3x - 1)(x + 2)$, contain opposite signs because the sign on the last term of the trinomial is ________.

$-;\ +$

Because $3x$ times 1 added to -2 times x gives the middle term, x.

311. $3x^2 + x - 2 = (3x\_\_\_2)(x\_\_\_1)$, as can be determined by consulting the middle term, x, of the trinomial.

Yes

The check always clears things up.

312. Does $(3x - 2)(x + 1) = 3x^2 + x - 2$?

opposite

The last term is negative.

313. If the binomial factors of $2x^2 - 5x - 7$ exist, they must contain (like/opposite) signs.

$-$; $+$

Because $2x - 7x = -5x$.

314. $2x^2 - 5x - 7 = (2x\_\_\_ 7)(x\_\_\_ 1)$, as can be determined by consulting the middle term of the trinomial, $-5x$.

Yes

315. Does $(2x - 7)(x + 1) = 2x^2 - 5x - 7$?

$+$; $-$

Check it.

316. $3x^2 - 5x - 2 = (3x\_\_\_ 1)(x\_\_\_ 2)$.

1; 3

317. $2x^2 - 5x - 3 = (2x + \_\_\_)(x - \_\_\_)$.

$(5x - 2)(x + 1)$

318. $5x^2 + 3x - 2 = (\_\_\_\_\_\_)(\_\_\_\_\_\_)$.

$(5x + 1)(x - 2)$

319. $5x^2 - 9x - 2 = (\_\_\_\_\_\_)(\_\_\_\_\_\_)$.

$(5x - 1)(x + 2)$

320. Factor $5x^2 + 9x - 2$.

$(2x + 7)(x - 1)$

Keep checking all factored forms.

321. Factor $2x^2 + 5x - 7$.

$(2x + 1)(x - 7)$

322. $2x^2 - 13x - 7 = (\_\_\_\_\_\_)(\_\_\_\_\_\_)$.

$(7x - 3)(x + 1)$

323. Factor $7x^2 + 4x - 3$.

$(7x + 3)(x - 1)$

It would be hard to see the factors of $-4x - 3 + 7x^2$.

324. To factor $-4x - 3 + 7x^2$, the trinomial is first written in standard form $7x^2 - 4x - 3$, which factors to ________________ .

$(5x - 1)(x + 3)$

325. Factor the trinomial $-3 + 14x + 5x^2$. (First write the trinomial in standard form.)

$(5x - 3)(x + 1)$

Did you first write the trinomial in standard form?

326. Factor $2x - 3 + 5x^2$.

$(5x + 1)(x - 3)$

327. Factor $5x^2 - 3 - 14x$.

$(3x + 1)(3x - 1)$

328. The difference of two squares $(a^2 - b^2)$ always factors to the form $(a + b)(a - b)$. Therefore, since $9x^2 - 1 = (3x)^2 - 1$, $9x^2 - 1 =$ (__________)(__________).

$(4x + 3)(4x - 3)$

329. $16x^2 - 9 =$ (__________)(__________).

$(6x + 5)(6x - 5)$

330. $36x^2 - 25 =$ (__________)(__________).

$(8x + 3)(8x - 3)$

331. $64x^2 - 9 =$ (__________)(__________).

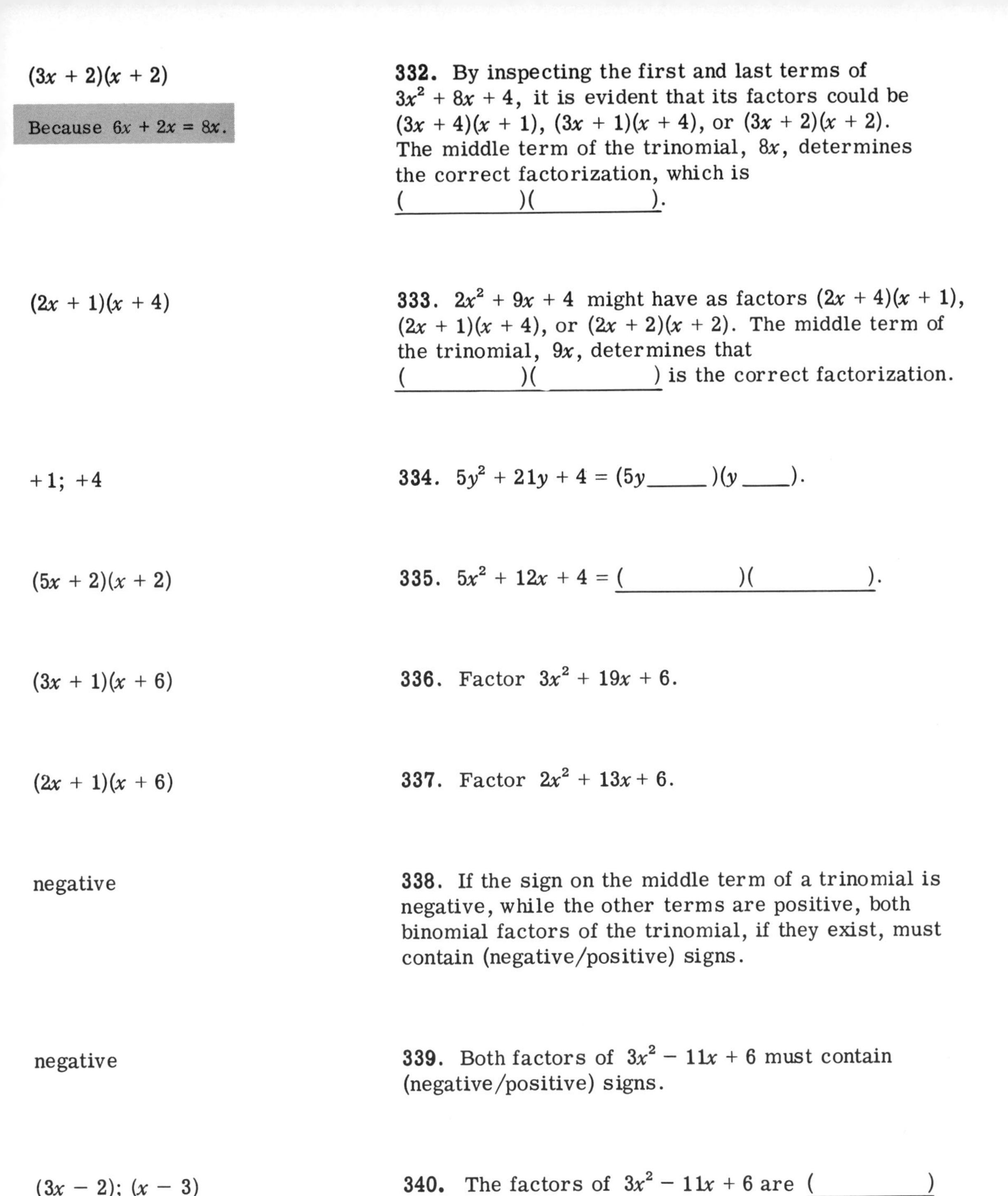

$(3x + 2)(x + 2)$

Because $6x + 2x = 8x$.

332. By inspecting the first and last terms of $3x^2 + 8x + 4$, it is evident that its factors could be $(3x + 4)(x + 1)$, $(3x + 1)(x + 4)$, or $(3x + 2)(x + 2)$. The middle term of the trinomial, $8x$, determines the correct factorization, which is (__________)(__________).

$(2x + 1)(x + 4)$

333. $2x^2 + 9x + 4$ might have as factors $(2x + 4)(x + 1)$, $(2x + 1)(x + 4)$, or $(2x + 2)(x + 2)$. The middle term of the trinomial, $9x$, determines that (__________)(__________) is the correct factorization.

$+1$; $+4$

334. $5y^2 + 21y + 4 = (5y$______$)(y$______$)$.

$(5x + 2)(x + 2)$

335. $5x^2 + 12x + 4 =$ (__________)(__________).

$(3x + 1)(x + 6)$

336. Factor $3x^2 + 19x + 6$.

$(2x + 1)(x + 6)$

337. Factor $2x^2 + 13x + 6$.

negative

338. If the sign on the middle term of a trinomial is negative, while the other terms are positive, both binomial factors of the trinomial, if they exist, must contain (negative/positive) signs.

negative

339. Both factors of $3x^2 - 11x + 6$ must contain (negative/positive) signs.

$(3x - 2)$; $(x - 3)$

340. The factors of $3x^2 - 11x + 6$ are (__________) and (__________).

$(5x - 2)(x - 2)$

341. Factor $5x^2 - 12x + 4$.

$(3x - 1)(x - 6)$

342. $3x^2 - 19x + 6 = (\underline{\qquad\qquad})(\underline{\qquad\qquad})$.

$(5x - 2)(x - 4)$

343. $5x^2 - 22x + 8 = (\underline{\qquad\qquad})(\underline{\qquad\qquad})$.

opposite

344. If the last term of a trinomial is negative while the first term is positive, for example $3x^2 + 4x - 4$, the signs in the factors of the trinomial, if they exist, must be (like/opposite).

opposite

345. The factors of $3x^2 - x - 4$, if they exist, must contain (like/opposite) signs.

$-$; $+$

Because $3x - 4x = -x$.

346. $3x^2 - x - 4 = (3x\_\_\_4)(x\_\_\_1)$.

$+$; $-$

347. $2x^2 - x - 6 = (2x\_\_\_3)(x\_\_\_2)$.

$-$; $+$

348. $2x^2 + x - 6 = (2x\_\_\_3)(x\_\_\_2)$.

$+$; $-$

Keep checking.

349. $3x^2 - 11x - 4 = (3x\_\_\_1)(x\_\_\_4)$.

$-$; $+$

350. $3x^2 + 11x - 4 = (3x\_\_\_1)(x\_\_\_4)$.

$+$; $-$

351. $5x^2 - 8x - 4 = (5x\_\_\_2)(x\_\_\_2)$.

$(2x + 3)(x - 2)$

352. Factor $2x^2 - x - 6$.

$(7x + 2)(x - 3)$

353. $7x^2 - 19x - 6 = (\underline{\qquad\quad})(\underline{\qquad\quad})$.

$(5x - 4)(x + 2)$

354. Factor $5x^2 + 6x - 8$.

$(x - 6)(2x + 3)$

355. Factor $2x^2 - 9x - 18$.

$(x - 6)(5x + 3)$

356. Factor $5x^2 - 27x - 18$.

5; 3

Because $9x + 10x = 19x$.

357. If both the first and last terms in a trinomial can be factored in more than one way, the possible factors of the trinomial increase in number. Observing the first and last term only, $6x^2 + 19x + 15$ might have as factors $(3x + 1)(2x + 15)$, $(3x + 15)(2x + 1)$, $(3x + 3)(2x + 5)$, $(3x + 5)(2x + 3)$, $(6x + 1)(x + 15)$, $(6x + 15)(x + 1)$, $(6x + 3)(x + 5)$, or $(6x + 5)(x + 3)$. The middle term of the trinomial, $19x$, determines that $(3x + \underline{\quad})(2x + \underline{\quad})$ is the correct factorization.

$(4x + 3)(x + 2)$

Because $8x + 3x = 11x$.

358. Possible factors of $4x^2 + 11x + 6$ are $(4x + 1)(x + 6)$, $(4x + 6)(x + 1)$, $(2x + 6)(2x + 1)$, $(4x + 3)(x + 2)$, $(4x + 2)(x + 3)$, and $(2x + 2)(2x + 3)$. The middle term of the trinomial, $11x$, determines that the correct factorization is $(\underline{\qquad\quad})(\underline{\qquad\quad})$.

$(x + 4)$

Check it.

359. The most efficient way to factor trinomials such as $6x^2 + 25x + 4$ is by mentally applying trial and error to the various possibilities. The factors of $6x^2 + 25x + 4$ are $(6x + 1)(\underline{\qquad\quad})$.

$(2x + 3)$

360. $6x^2 + 13x + 6 = (3x + 2)(\underline{\qquad\quad})$.

$+5$; $+3$

361. $6x^2 + 21x + 15 = (2x$____$)(3x$____$)$.

$(4x + 3)(x + 5)$

362. Factor $4x^2 + 23x + 15$.

$(2x + 7)(2x + 5)$

363. Factor $4x^2 + 24x + 35$.

$(6x + 1)(x + 6)$

364. $6x^2 + 37x + 6 =$ (__________)(__________).

$(2x + 1)(x + 6)$

365. Factor $2x^2 + 13x + 6$.

negative

366. Since the sign on the middle term of $4x^2 - 11x + 6$ is negative, while the sign on the last term is positive, the signs in both factors of $4x^2 - 11x + 6$ are (negative/positive).

$-$; $-$

367. $4x^2 - 11x + 6 = (4x$____$3)(x$____$2)$.

-1; -4

368. $6x^2 - 11x + 4 = (2x$____$)(3x$____$)$.

$(3x - 2)(2x - 5)$

369. Factor $6x^2 - 19x + 10$.

$(2x - 2)(3x - 5)$

370. $6x^2 - 16x + 10 =$ (__________)(__________).

$(4x - 3)(2x - 5)$

371. Factor $8x^2 - 26x + 15$.

opposite

Otherwise, the last term would not be negative. Remember?

372. Because the last term of $4x^2 + 4x - 15$ is negative and the first term is positive, the binomial factors of $4x^2 + 4x - 15$, if they exist, contain __________________ signs.

$-; +$

373. $4x^2 + 4x - 15 = (2x$ ___ $3)(2x$ ___ $5)$.

$2x + 5$

374. $6x^2 + 11x - 10 = (3x - 2)($_________$)$.

$(4x - 3)(x + 2)$

375. $4x^2 + 5x - 6 =$ (_________)(_________).

$(3x + 2)(2x - 5)$

376. Factor $6x^2 - 11x - 10$.

$6x^2 + x - 35$

Remember, descending powers of the variable.

377. Before trying to factor $x - 35 + 6x^2$, the expression should be written in standard form, which is ____________________ .

$(2x + 5)(3x - 7)$

378. Factor $6x^2 + x - 35$.

$(4x - 5)(2x + 3)$

The standard form of the trinomial is $8x^2 + 2x - 15$.

379. First write $2x - 15 + 8x^2$ in standard form and then factor.

$(3x - 1)(3x + 4)$

380. Factor $9x - 4 + 9x^2$.

Remark. Sometimes, in factoring trinomials, you have to consider two different kinds of factors in the same trinomial. You might find that you have both a monomial factor and also two binomial factors. Factoring such trinomials does not involve anything that you haven't already covered; you just have to do the factoring in two steps.

3

381. If the terms of a trinomial contain a common monomial factor, the monomial factor should be factored from the trinomial before trying to factor the trinomial into two binomials. The trinomial $3x^2 + 12x + 12$ contains the monomial factor_____.

$(x + 2)$

382. To factor $3x^2 + 12x + 12$ completely, it is best to write the trinomial as $3(x^2 + 4x + 4)$ first, and then as $3(x + 2)(\underline{\qquad\qquad})$.

2

383. Each term of the trinomial $2x^2 - 8x - 10$ contains the monomial factor_____.

$2(x - 5)(x + 1)$

384. To factor $2x^2 - 8x - 10$ completely, it is best to write $2(x^2 - 4x - 5)$ first, and then $\underline{2(\qquad\qquad)(\qquad\qquad)}$.

3

385. Each term of the trinomial $27y^2 - 9y - 6$ contains the monomial factor_____.

$9y^2 - 3y - 2;$
$3(3y + 1)(3y - 2)$

386. $27y^2 - 9y - 6 = 3(\underline{\qquad\qquad\qquad\qquad})$
$= \underline{3(\qquad\qquad)(\qquad\qquad)}$.

$3(3x + 1)(x - 2)$

First factor out the monomial 3.

387. Factor $9x^2 - 15x - 6$ completely.

x

388. The monomial factor common to each term of $x^3 - 2x^2 + x$ is ____ .

$x(x - 1)(x - 1)$

You can write $x(x - 1)^2$ if you wish.

389. Factor $x^3 - 2x^2 + x$ completely.

$2x(3x + 1)(x + 1)$

You first have $2x(3x^2 + 4x + 1)$.

390. Factor $6x^3 + 8x^2 + 2x$ completely.

$3(3x + 1)(2x - 1)$

You first have $3(6x^2 - x - 1)$.

391. In completely factored form,
$18x^2 - 3x - 3 =$ 3(   )(   ).

$3x(x - 1)(x + 1)$

392. $3x^3 - 3x = 3x(x^2 - 1)$
= 3x(   )(   ).

$3(a - 5)(a + 5)$

393. $3a^2 - 75 = 3(a^2 - 25)$
= 3(   )(   ).

$2(x - 2)(x + 2)$

394. Factor $2x^2 - 8$ completely.

$3(y - 2)(y + 2)$

395. Factor $3y^2 - 12$ completely.

$x^2(x - 1)(x + 1)$

396. In completely factored form,
$x^4 - x^2 =$ ________________ .

$(a + b)$

397. If more than one variable occurs in a trinomial, the factoring process does not change. Thus,
$x^2 + 3xy + 2y^2 = (x + 2y)(x + y)$.
$a^2 + 4ab + 3b^2 = (a + 3b)(\underline{\hspace{2cm}})$.

$(2x + y)(x + 3y)$

398. $2x^2 + 7xy + 3y^2 = (\underline{\hspace{2cm}})(\underline{\hspace{2cm}})$.

$(2a - b)(a + 3b)$

399. Factor $2a^2 + 5ab - 3b^2$.

$(2a - 5b)(2a + 5b)$

400. $4a^2 - 25b^2 = (\underline{\hspace{2cm}})(\underline{\hspace{2cm}})$.

$3a(4b + a)(b + a)$

401. $12ab^2 + 15a^2b + 3a^3 = 3a(4b^2 + 5ab + a^2)$
$= 3a(\underline{\hspace{2cm}})(\underline{\hspace{2cm}})$.

$3b(3a - b)(3a + 4b)$

Did you first write
$3b(9a^2 + 9ab - 4b^2)$?

402. $27a^2b + 27ab^2 - 12b^3 = \underline{\hspace{4cm}}$.

$3x(y - 2b)(y + 2b)$

You first write
$3x(y^2 - 4b^2)$.

403. $3xy^2 - 12xb^2 = \underline{\hspace{4cm}}$.

$7x$

Because
$(1 + x)(1 - 7x) = 1 - 6x - 7x^2$.

404. The trinomial $1 - 6x - 7x^2$ is not in standard form but can be factored as it stands. The factors are $(1 + x)(1 - \underline{\hspace{1cm}})$.

$(2 - x)$

Because
$(3 - x)(2 - x) = 6 - 5x + x^2$.

405. The trinomial $6 - 5x + x^2$ can be factored as $(3 - x)(\underline{\hspace{2cm}})$.

$(1 - x)$

406. The trinomial $1 + 2x - 3x^2$ can be factored as $(1 + 3x)(\_\_\_\_\_\_\_\_\_\_)$.

$(1 + 3x)(1 - x)$

407. With some trinomials, where the coefficient of x^2 is negative, it is helpful to write the trinomial in the reverse of standard form before factoring. Thus, $-3x^2 + 2x + 1$ can be written $1 + 2x - 3x^2$, which factors to __________________.

$(3 - x)$

408. $-x^2 - 4x + 21 = 21 - 4x - x^2$

$$= (7 + x)(\_\_\_\_\_\_\_\_\_\_).$$

$(9 - y)(2 + y)$

409. Factor $18 + 7y - y^2$.

$(3 + x)(2 - x)$

410. Factor $-x^2 - x + 6$.

$(5 - 3x)(5 + 3x)$

411. $25 - 9x^2 = (\_\_\_\_\_\_\_\_\_\_)(\_\_\_\_\_\_\_\_\_\_)$. *

Remark. You can now use the distributive law to multiply polynomials and also to factor polynomials. Both of these processes have applications in the solution of equations. Later, when you study the solution of quadratic equations, you will need to know how to factor polynomials. Right now, however, we are going to apply the distributive law to help solve some first-degree equations that involve parentheses.

equivalent

412. Recall that equivalent equations are equations having exactly the same solutions. Because 3 is the only solution of each, $x - 1 = 2$ and $x = 3$ are __________________ equations.

* See Exercise 5, page 62, for additional practice.

distributive

413. An equation containing parentheses, such as $2(x + 5) = 16$, can be transformed to an equivalent equation that does not contain parentheses, $2x + 10 = 16$, by applying the_________________ law.

3

414. Solve $2x + 10 = 16$ for x. The solution is_____.

2

415. The equation $6 = 2(2x - 1)$ is equivalent to the equation $6 = 4x - 2$. Solve this equation for x. The solution is_____.

3

416. Solve $5b + 10(8 - b) = 65$, for b. The solution is_____.

6

417. If the distributive law is applied to both members of $4(y - 1) = 5(y - 2)$, the result is $4y - 4 = 5y - 10$. Solve this equation for y. The solution is_____.

-2

418. Solve $3(7 + 2x) = 30 + 7(x - 1)$ for x. The solution is_____.

3

419. Solve $5a - 4(1 - a) = 11 - 6(a - 5)$ for a. The solution is_____.

-6

420. Solve $7 - (x - 2) = 3(11 + x)$ for x. The solution is_____.

Remark. Having learned to solve first-degree equations that involve parentheses, it is time to look at some specific applications of such equations. Remember, as you work through the frames ahead, that parentheses are used to group terms that are to be viewed as a single number.

$5(x + 3)$

421. If x represents a number, then $x + 3$ represents a number that is three greater than x. Two times the larger number can be represented by $2(x + 3)$; five times the larger number can be represented by__________.

$x + 4$; $5(x + 4)$

422. One number is four more than a second number. If x represents the smaller number, then the larger number would be represented by__________, and 5 times the larger number would be represented by__________.

$n - 6$; $7(n - 6)$

423. One number is six less than a second number. If n represents the larger number, then the smaller number would be represented by__________, and seven times the smaller number would be represented by__________.

$x - 2$; $4(x - 2)$

424. One number is two less than a second number. If x represents the larger number then __________ represents the smaller, and four times the smaller would be represented by__________.

$x + 4(x - 2) = 17$

425. If x represents a number, and $x - 2$ represents a smaller number, then the statement, "The larger plus four times the smaller equals 17," can be represented by the equation__________ = _____.

5

426. The solution of $x + 4(x - 2) = 17$ is_____.

5; 3

If x is 5 then $x - 2$ is 3.

427. Since 5 is the solution of $x + 4(x - 2) = 17$, the larger number in Frame 425 is_____ and the smaller number is_____.

$6;\ 9$

You first have $4x + 2(x + 3) = 42$ where x represents the smaller number.

428. One number is three more than a second number. If four times the smaller plus twice the larger equals 42 the numbers are_____ and_____ .

$2(l - 6)$

429. If the width of a rectangle is 6 feet less than its length, l, two times the width would be represented in terms of the length by__________ .

$2l + 2(l - 6) = 44$

430. The length of a rectangle is represented by l and the width is represented by $l - 6$. The statement that the perimeter (two times its length plus two times its width) equals 44 feet can be represented by the equation__________________ = _______.

14

431. The solution of $2l + 2(l - 6) = 44$ is______.

14 ft.; 8 ft.

432. Since 14 is the solution of $2l + 2(l - 6) = 44$, the length of the rectangle in Frame 430 is__________ and the width__________.

$24 - y$

The whole length, 24, minus the longer piece, y.

433. A 24-foot board is divided into two parts. If y represents the length of the longer piece, the length of the shorter piece would be represented in terms of y by____________ .

$3(24 - y)$

434. If y represents the length of the longer piece of a board and $24 - y$ represents the shorter piece, then three times the shorter piece would be represented by_____________.

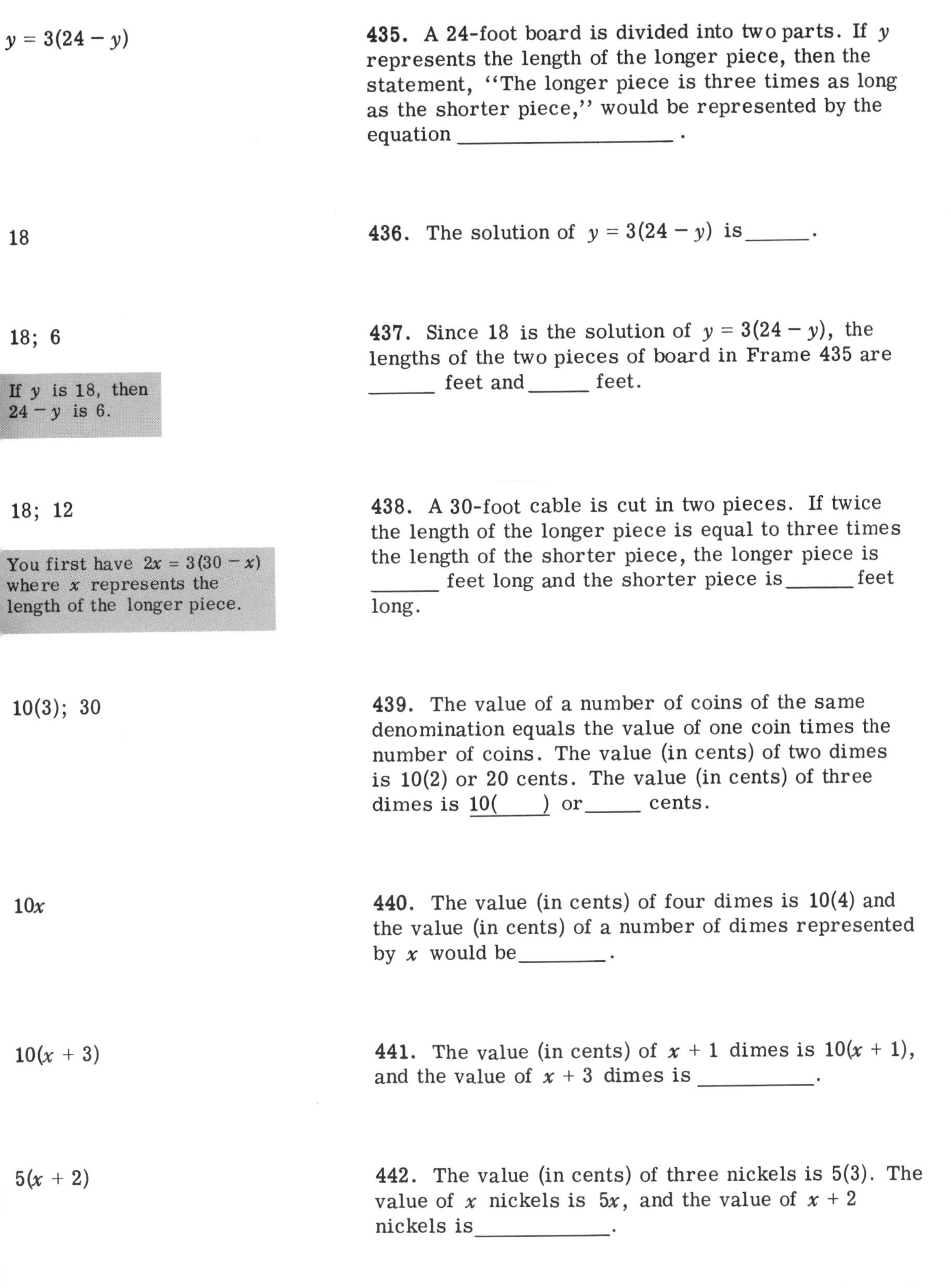

$y = 3(24 - y)$

435. A 24-foot board is divided into two parts. If y represents the length of the longer piece, then the statement, "The longer piece is three times as long as the shorter piece," would be represented by the equation ________________ .

18

436. The solution of $y = 3(24 - y)$ is______.

18; 6

If y is 18, then $24 - y$ is 6.

437. Since 18 is the solution of $y = 3(24 - y)$, the lengths of the two pieces of board in Frame 435 are ______ feet and______ feet.

18; 12

You first have $2x = 3(30 - x)$ where x represents the length of the longer piece.

438. A 30-foot cable is cut in two pieces. If twice the length of the longer piece is equal to three times the length of the shorter piece, the longer piece is ______ feet long and the shorter piece is______ feet long.

10(3); 30

439. The value of a number of coins of the same denomination equals the value of one coin times the number of coins. The value (in cents) of two dimes is 10(2) or 20 cents. The value (in cents) of three dimes is 10(____) or_____ cents.

$10x$

440. The value (in cents) of four dimes is 10(4) and the value (in cents) of a number of dimes represented by x would be________.

$10(x + 3)$

441. The value (in cents) of $x + 1$ dimes is $10(x + 1)$, and the value of $x + 3$ dimes is __________.

$5(x + 2)$

442. The value (in cents) of three nickels is 5(3). The value of x nickels is $5x$, and the value of $x + 2$ nickels is____________.

$25(y - 3)$

443. The value (in cents) of five quarters is $25(5)$, the value of y quarters is $25y$, and the value of $y - 3$ quarters is ____________.

$x + 4$

444. In a collection of coins there are four more dimes than quarters. If x represents the number of quarters, the number of dimes would be represented in terms of x by__________.

$25x$; $10(x + 4)$

445. If x represents the number of quarters in a collection of coins and $x + 4$ represents the number of dimes then, in terms of x, the value (in cents) of the quarters would be represented by_________ and the value of the dimes would be represented by ____________.

$25x + 10(x + 4)$

The value of the quarters plus the value of the dimes.

446. If $25x$ represents the value (in cents) of the quarters in a collection of coins and $10(x + 4)$ represents the value of the dimes, the total value of the collection (in cents) would be represented in terms of x by_______+_______________, where x represents the number of quarters and $x + 4$ represents the number of dimes.

7

447. If the total value of the collection of dimes and quarters in Frame 444 was $2.85 or 285 cents, then $25x + 10(x + 4) = 285$. The solution of this equation is_____.

7; 11

If x is 7, then $x + 4$ is 11.

448. Since 7 is the solution of $25x + 10(x + 4) = 285$, the collection of coins in Frame 444 contains_______ quarters and________dimes.

$5x + 10(x + 3)$

449. A collection of coins containing only nickels and dimes contained three more dimes than nickels. If x represents the number of nickels, the value of the nickels (in cents) is $5x$, the value of the dimes (in cents) is $10(x + 3)$, and the value of the entire collection (in cents) in terms of x is ______________.

180

The value of the nickels plus the value of the dimes equals the value of the total collection.

450. If the value of the collection in Frame 449 is \$1.80 or 180 cents, then $5x + 10(x + 3) =$ __________ .

10

451. The solution of $5x + 10(x + 3) = 180$ is__________.

10; 13

If x is 10, then $x + 3$ is 13.

452. Since 10 is the solution of $5x + 10(x + 3) = 180$, the collection of coins in Frame 449 contains______ nickels and_________dimes.

3; 7

You first have $25x + 10(x + 4) = 145$, where x represents the number of quarters.

453. A man had \$1.45 in change consisting of quarters and dimes only. If he had four more dimes than quarters, he had_____quarters and_____dimes.

$40x + 90(300 - x)$

The value of childrens' tickets plus the value of adults' tickets.

454. Three hundred tickets were sold at a baseball game. If x represents the number of children's tickets at 40 cents each and $300 - x$ represents the number of adults' tickets at 90 cents each, the total receipts for the game (in cents) can be represented in terms of x by________________________.

140

455. If the total receipts of the game in Frame 454 was \$200 or 20,000 cents then $40x + 90(300 - x) = 20{,}000$. The solution of this equation is__________.

140; 160

If x is 140 then $300 - x$ is 160.

456. Since 140 is the solution of $40x + 90(300 - x) = 20{,}000$, there were_________ children's tickets and______ adults' tickets sold.

600; 400

You first have $80x + 180(1000 - x) = 120000$, where x represents the number of children's tickets; or $80(1000 - x) + 180x = 120000$, where x represents the number of adult tickets.

457. Adults paid $1.80 each and children paid $.80 each for their tickets to a football game. If one thousand tickets were sold and the receipts were $1200, there were__________ children's tickets and __________ adults' tickets sold.

$12(50) + 30x$

Or $600 + 30x$.

458. The value of 50 pounds of coarse powder that sells for 12 cents a pound is 12(50) cents. The value of x pounds of fine powder that sells for 30 cents a pound is $30x$. The value of a mixture of the two powders in terms of x is ______________ cents.

$20(50 + x)$

459. The total weight of the mixture of fine and coarse powder in Frame 458 can be represented in terms of x by $50 + x$ pounds. If the mixture sells for 20 cents per pound, the value of the mixture is ___________________cents.

$20(50 + x)$

460. Since the value of the mixture in Frame 458 can be represented by either $12(50) + 30x$ or by $20(50 + x)$, then $12(50) + 30x =$________________.

40

461. The solution of $12(50) + 30x = 20(50 + x) =$______.

40

462. Since the solution of $12(50) + 30x = 20(50 + x)$ is 40, the number of pounds of fine powder used in the mixture in Frame 458 is _____.

50

You first have $30(100) + 24(x) = 28(x + 100)$, where x represents the number of pounds of coarse powder.

463. A man makes a mixture of 100 pounds of fine powder worth 30 cents a pound and some coarse powder worth 24 cents a pound. If he wishes the mixture to sell for 28 cents a pound, he must use ________ pounds of coarse powder.*

Remark. While there are many other situations in which equations involving parentheses prove helpful, the ones you have learned to deal with are typical, and are all we propose to examine here.

The rest of this unit is a review, and the last sequence of frames will give you an opportunity to obtain an overview of the unit in a very short time. Although you should not anticipate learning anything you have not learned to this point, the review will highlight the main ideas in a brief way and will reinforce what you have learned. Since these last frames do not include a review of applications in the form of word problems, if you feel in need of another look at these, it might prove worthwhile for you at the end of the unit to reread quickly Frames 421 to 463.

$4a + 4b$

464. The principle that is stated symbolically $a(b + c) = ab + ac$ is called the distributive law. The distributive law guarantees that $4(a + b)$ equals ____________.

$x - 16$

465. To simplify $3(x - 4) - 2(x + 2)$ means to apply the distributive law and then add any like terms. The result of doing this is________.

$2(x - 3y)$

466. The distributive law in the form $ab + ac = a(b + c)$ permits writing a polynomial as a product. Thus, the polynomial $2x - 6y$ can be written as the product ____________.

* See Exercise 6, page 63, for additional practice.

factored

467. The process of writing a polynomial as a product of two or more factors is called factoring. If the terms within the parentheses of an expression of the form $a(b + c)$ contain no common monomial factors, the expression is said to be completely

____________________.

$2x(4x^2 - x + 2)$

468. $8x^3 - 2x^2 + 4x$ can be factored as

____________________.

$ac + ad + bc + bd$

469. Applying the distributive law twice permits $(a + b)(c + d)$ to be written as the polynomial

____________________________.

$3x^2 + 22x - 16$

470. Write $(3x - 2)(x + 8)$ as a trinomial.

$4x^2 - 4x + 1$

471. Write $(2x - 1)^2$ as a trinomial.

$x^2 - 4$

472. The product of the sum and difference of two numbers is always the difference of the squares of the numbers. Thus, $(x - 2)(x + 2) =$ __________.

$a(x^2 + x + 6)$

473. Factoring a trinomial is the process of writing a sum as a product. When the sum $ax^2 + ax + 6a$ is factored, it is written as the product

____________________.

$(x + 6)(x + 1)$

474. $x^2 + 7x + 6$ can be written as the product (__________)(__________).

$(x - 6)(x + 2)$

475. Factor $x^2 - 4x - 12$.

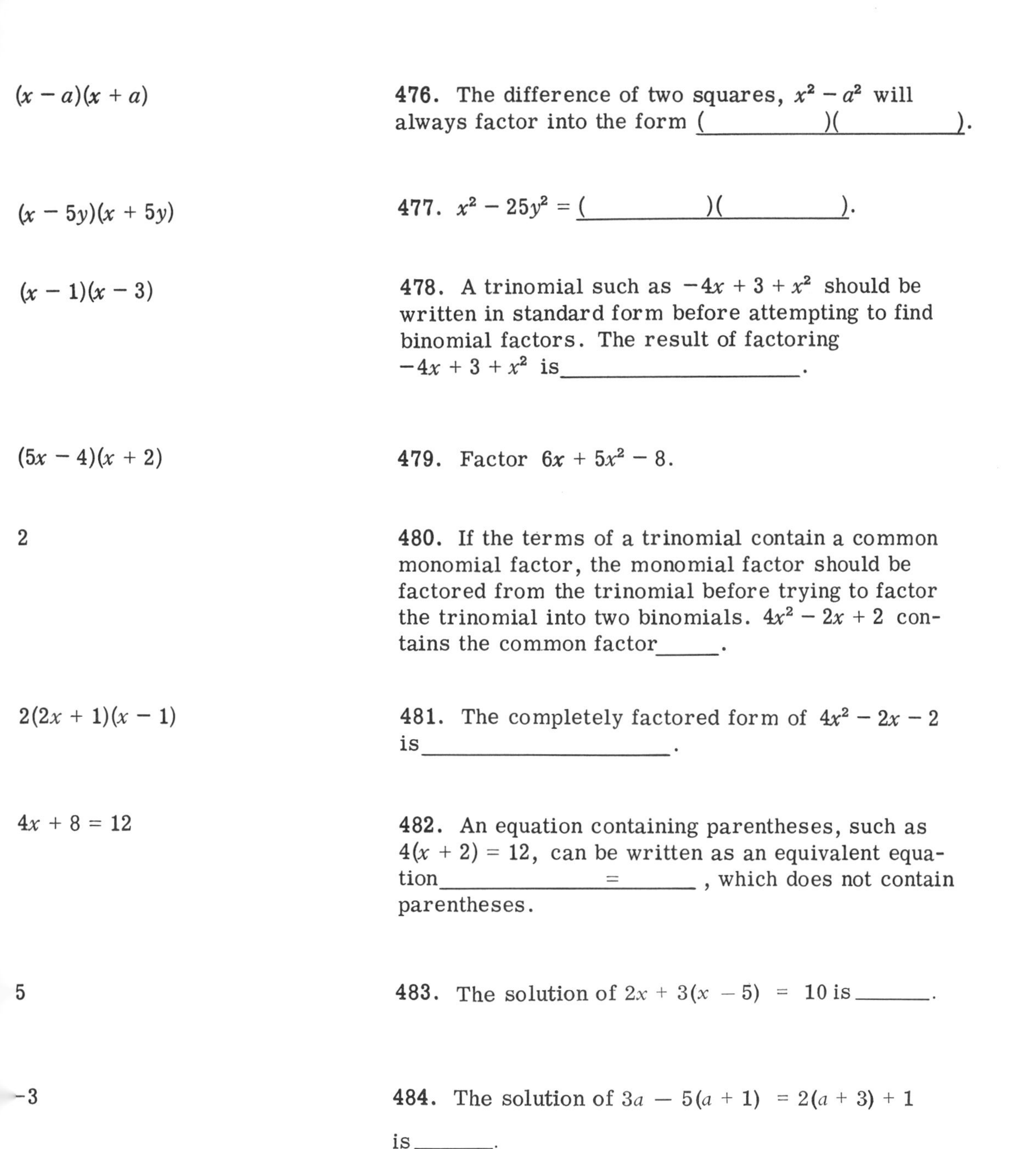

$(x - a)(x + a)$

476. The difference of two squares, $x^2 - a^2$ will always factor into the form (________)(________).

$(x - 5y)(x + 5y)$

477. $x^2 - 25y^2 =$ (________)(________).

$(x - 1)(x - 3)$

478. A trinomial such as $-4x + 3 + x^2$ should be written in standard form before attempting to find binomial factors. The result of factoring $-4x + 3 + x^2$ is________________.

$(5x - 4)(x + 2)$

479. Factor $6x + 5x^2 - 8$.

2

480. If the terms of a trinomial contain a common monomial factor, the monomial factor should be factored from the trinomial before trying to factor the trinomial into two binomials. $4x^2 - 2x + 2$ contains the common factor_____.

$2(2x + 1)(x - 1)$

481. The completely factored form of $4x^2 - 2x - 2$ is________________.

$4x + 8 = 12$

482. An equation containing parentheses, such as $4(x + 2) = 12$, can be written as an equivalent equation__________ = ______, which does not contain parentheses.

5

483. The solution of $2x + 3(x - 5) = 10$ is______.

-3

484. The solution of $3a - 5(a + 1) = 2(a + 3) + 1$ is______.

Remark. This completes the unit on products and factors. To see what you have learned take one form of the self-evaluation test on pages 64-65. If you completed one form before starting this unit, use the alternate form now.

EXERCISES AND ANSWERS

Exercise 1. If you have difficulty with this exercise, reenter the program at Frame 1.

Write each expression without using parentheses and then simplify whenever it is possible to do so.

1. $2(x+1)$ **2.** $5(x-3)$ **3.** $3(3x^2-2x+1)$ **4.** $-4(2x^2+x-3)$ **5.** $x(x^2+x-4)$

6. $-3x(2x^2-x+3)$ **7.** $2(x^2-x)+3(x+1)$ **8.** $5(x-2)-2(x^2+3)$

9. $4(x^2-2)-3(2x^2-1)$ **10.** $x(x+2)-2(x^2-x)$ **11.** $4(3x^2+1)-2x(x+1)$

12. $5x(x-2)-x(x-3)$ **13.** $2(x^2-4x+3)+4(2x^2+x)$ **14.** $3(2x^2-2)-5(x^2-x-1)$

15. $(x^2+2x-1)+(x+2)-(x^2-3x)$ **16.** $(4x-2)-(x^2-3x+2)-(x+2)$

17. $(x^2-3x+2)-(4x^2-x)-(x+1)$ **18.** $2x-(x^2-1)+(3x^2+2x+1)$

Answers

1. $2x+2$ **2.** $5x-15$ **3.** $9x^2-6x+3$ **4.** $-8x^2-4x+12$ **5.** x^3+x^2-4x

6. $-6x^3+3x^2-9x$ **7.** $2x^2+x+3$ **8.** $-2x^2+5x-16$ **9.** $-2x^2-5$

10. $-x^2+4x$ **11.** $10x^2-2x+4$ **12.** $4x^2-7x$ **13.** $10x^2-4x+6$ **14.** x^2+5x-

15. $6x+1$ **16.** $-x^2+6x-6$ **17.** $-3x^2-3x+1$ **18.** $2x^2+4x+2$

Exercise 2. If you have difficulty with this exercise, reenter the program at Frame 52.

Write each expression in factored form.

1. $2x+6$ **2.** $3x+9$ **3.** $3x^2-6x+9$ **4.** $4x^2+4x-2$ **5.** $2x^2-4x+4$

6. $4x^2-12x-4$ **7.** $2ax+4ay-6az$ **8.** $4bx-2by+2bz$ **9.** $3ax-3ay+6az$

10. $8ax+6ay-4az$ **11.** $-3x^2+9x$ **12.** $-4x^2-6x$ **13.** $-2x^2+4x$

14. $-5x^2+5x-10$ **15.** $6x^2y+3xy$ **16.** $12xy-6xy^2$ **17.** x^2y+xy^2

18. $4xy^2+2xy$ **19.** $15x^2y+12xy^2$ **20.** $8xy^2-20x^2y$ **21.** $8x^2y+20xy$

22. $-4x^2y-18xy^2$ **23.** $18x^2y+6xy-4y$ **24.** $4x^2y-6xy-2x$

25. $2x^2y+xy^2-3xy$ **26.** $10x^2y^2-5xy-5x^2y$ **27.** $-x^3y+x^2y^2+xy^3$

28. $-xy^3-x^2y-x^3y$ **29.** $6x^3y^2-2x^2y^2+4x^2y^3$ **30.** $15xy^3-12x^2y^2+9x^3y$

Answers

1. $2(x + 3)$ 2. $3(x + 3)$ 3. $3(x^2 - 2x + 3)$ 4. $2(2x^2 + 2x - 1)$ 5. $2(x^2 - 2x + 2)$

6. $4(x^2 - 3x - 1)$ 7. $2a(x + 2y - 3z)$ 8. $2b(2x - y + z)$ 9. $3a(x - y + 2z)$

10. $2a(4x + 3y - 2z)$ 11. $-3x(x - 3)$ 12. $-2x(2x + 3)$ 13. $-2x(x - 2)$

14. $-5(x^2 - x + 2)$ 15. $3xy(2x + 1)$ 16. $-6xy(y - 2)$ 17. $xy(x + y)$

18. $2xy(2y + 1)$ 19. $3xy(5x + 4y)$ 20. $4xy(2y - 5x)$ 21. $4xy(2x + 5)$

22. $-2xy(2x + 9y)$ 23. $2y(9x^2 + 3x - 2)$ 24. $2x(2xy - 3y - 1)$ 25. $xy(2x + y - 3)$

26. $5xy(2xy - x - 1)$ 27. $-xy(x^2 - xy - y^2)$ 28. $-xy(y^2 + x + x^2)$

29. $2x^2y^2(3x - 1 + 2y)$ 30. $3xy(5y^2 - 4xy + 3x^2)$

Exercise 3. If you have difficulty with this exercise, reenter the program at Frame 127.

Write each expression without using parentheses and then simplify whenever it is possible to do so.

1. $(x + 2)(x + 3)$	2. $(x - 5)(x + 2)$	3. $(x - 7)(x - 3)$
4. $(x + 8)(x - 2)$	5. $(2x - 3)(x + 4)$	6. $(3x + 1)(3x - 2)$
7. $(x + 3y)(x - y)$	8. $(x + y)(2x - y)$	9. $(x + 4)^2$
10. $(2x - 1)^2$	11. $(x + 5)(x - 5)$	12. $(3x - 2)(3x + 2)$
13. $3(x - 1)(x + 2)$	14. $4(2x - 1)(x + 3)$	15. $5(x - 1)^2$
16. $2x(x + 2)^2$	17. $3(x - 4)(x + 4)$	18. $2(x + 6)(x - 6)$

Answers

1. $x^2 + 5x + 6$	2. $x^2 - 3x - 10$	3. $x^2 - 10x + 21$
4. $x^2 + 6x - 16$	5. $2x^2 + 5x - 12$	6. $9x^2 - 3x - 2$
7. $x^2 + 2xy - 3y^2$	8. $2x^2 + xy - y^2$	9. $x^2 + 8x + 16$
10. $4x^2 - 4x + 1$	11. $x^2 - 25$	12. $9x^2 - 4$
13. $3x^2 + 3x - 6$	14. $8x^2 + 20x - 12$	15. $5x^2 - 10x + 5$
16. $2x^3 + 8x^2 + 8x$	17. $3x^2 - 48$	18. $2x^2 - 72$

Exercise 4. If you have difficulty with this exercise, reenter the program of Frame 205.

Write each of the following in factored form.

1. $x^2 + 4x + 3$ **2.** $x^2 - 3x + 2$ **3.** $x^2 + 3x - 10$ **4.** $x^2 - x - 6$ **5.** $x^2 - 3x - 28$

6. $x^2 + 2x - 48$ **7.** $x^2 - 4$ **8.** $x^2 - 16y^2$ **9.** $x^2 - 4x + 4$ **10.** $x^2 - 6x + 9$

11. $-12 - x + x^2$ **12.** $-25 + x^2$

Answers

1. $(x + 3)(x + 1)$ **2.** $(x - 2)(x - 1)$ **3.** $(x + 5)(x - 2)$ **4.** $(x - 3)(x + 2)$

5. $(x + 4)(x - 7)$ **6.** $(x + 8)(x - 6)$ **7.** $(x + 2)(x - 2)$ **8.** $(x + 4y)(x - 4y)$

9. $(x - 2)^2$ **10.** $(x - 3)^2$ **11.** $(x - 4)(x + 3)$ **12.** $(x + 5)(x - 5)$

Exercise 5. If you have difficulty with this exercise, reenter the program at Frame 292.

Write each of the following in factored form.

1. $3x^2 + 11x + 6$ **2.** $3x^2 - 7x + 2$ **3.** $5x^2 - 13x - 6$ **4.** $2x^2 + 5x - 3$ **5.** $9x^2 - 4$

6. $16x^2 - 1$ **7.** $6x^2 + 11x + 3$ **8.** $4x^2 - 8x + 3$ **9.** $8x^2 + 2x - 1$ **10.** $6x^2 - 13x -$

11. $2x^2 + 6x + 4$ **12.** $3x^2 - 12x + 9$ **13.** $6x^2 - 24$ **14.** $x^3 - x$ **15.** $2x^2 + 3xy + y^2$

16. $3x^2 - 2xy - y^2$ **17.** $12 - 11x + 2x^2$ **18.** $-x^2 - x + 6$

Answers

1. $(3x + 2)(x + 3)$ **2.** $(3x - 1)(x - 2)$ **3.** $(5x + 2)(x - 3)$ **4.** $(2x - 1)(x + 3)$

5. $(3x + 2)(3x - 2)$ **6.** $(4x + 1)(4x - 1)$ **7.** $(3x + 1)(2x + 3)$ **8.** $(2x - 3)(2x - 1)$

9. $(4x - 1)(2x + 1)$ **10.** $(3x + 1)(2x - 5)$ **11.** $2(x + 2)(x + 1)$ **12.** $3(x - 3)(x - 1)$

13. $6(x + 2)(x - 2)$ **14.** $x(x + 1)(x - 1)$ **15.** $(2x + y)(x + y)$ **16.** $(3x + y)(x - y)$

17. $(2x - 3)(x - 4)$ **18.** $-1(x + 3)(x - 2)$

Exercise 6. If you have difficulty with this exercise, reenter the program at Frame 412.

1. a. Represent the number that is five times greater than a number represented by x.

   b. Represent the number that is three times the larger number.

2. a. Represent the number that is four less than a number represented by n.

   b. Represent the number that is six times the smaller number.

3. a. The length of a rectangle is 8 feet greater than its width, w. Represent the length in terms of w.

   b. Represent two times the length in terms of w.

   c. Find the dimensions of the rectangle if the perimeter is 40 feet.

4. a. The width of a rectangle is 4 feet less than its length, l. Represent the width in terms of l.

   b. Represent two times the width in terms of l.

   c. Find the dimensions of the rectangle if the perimeter is 44 feet.

5. a. A 45-foot rope is cut in two pieces. If x represents the length of the longer piece, represent four times the length of the shorter piece.

   b. Find the length of each piece of rope if the length of the longer piece equals four times the length of the shorter piece.

6. a. In a collection of coins there are six more quarters than dimes. If x represents the number of dimes, represent the value of the dimes.

   b. Represent the value of the quarters in terms of x.

   c. Find the number of dimes and quarters in the collection of coins if the total value is \$3.25.

7. a. In Problem 6, if y represents the number of quarters, represent the value of the dimes in terms of y.

   b. Write the equation using the conditions on y to express the fact that the total value of the collection is \$3.25.

   c. Find the number of each kind of coin using the equation you wrote in **b.**

8. How many of each kind of coin does he have if he has four more dimes than nickels?

Answers

1. a. $x + 5$ **b.** $3(x + 5)$ **2. a.** $n - 4$ **b.** $6(n - 4)$ **3. a.** $8 + w$ **b.** $2(8 + w)$

c. $6' \times 14'$ **4. a.** $L - 4$ **b.** $2(L - 4)$ **c.** $9' \times 13'$ **5. a.** $4(45 - x)$ **b.** 36 feet; 9 feet

6. a. $x(.10)$ **b.** $(x + 6)(.25)$ **c.** 5 dimes; 11 quarters **7. a.** $(y - 6)(.10)$

b. $(y - 6)(.10) + y(.25) = \$3.25$ **c.** 11 quarters; 5 dimes **8.** 23 dimes; 19 nickels

SELF-EVALUATION TEST, FORM A

1. When $2(x + 1) - 3(x - 1)$ is simplified, the result is _________ .
2. Write $a(a + 4)(a + 1)$ as a trinomial.
3. The complete factorization of $3y^2 + 5y + 2$ is _______________.
4. The solution of $3(2x + 1) = x + 8$ is_____.
5. The basic principle underlying all factoring is the_______________ law.
6. When $-5y(2y^2 + y - 1)$ is written as a trinomial, the result is __________________.
7. Completely factor $3x^3 + 12x^2 - 3x$.
8. Completely factor $x^2 - 6x + 8$.
9. Completely factor $a^2c - 9b^2c$.
10. Completely factor $4x^2 + 14xy + 6y^2$.
11. When $(3x + 1)^2$ is written as a trinomial, the result is_______________.
12. In a collection of coins there are six more dimes than quarters. If the total value of the dimes and the quarters in the collection is \$3.05, there are______ dimes and _____ quarters in the collection.

SELF-EVALUATION FORM TEST, FORM B

1. When $x(x - 3) - x^2$ is simplified, the result is______.
2. Write $b(b - 3)(b + 1)$ as a trinomial.
3. The complete factorization of $2x^2 + 5x - 3$ is_____________.
4. The solution of $3(x + 1) - 2 = 7$ is_____.
5. When $-3a(3a^2 - a + 1)$ is written as a trinomial the result is_______________.
6. Completely factor $2a^3 - 8a^2 + 2a$.
7. The basic principle underlying all factoring is the________________ law.
8. Completely factor $a^2 + 4ab + 3b^2$.
9. Completely factor $xy^2 - 4x$.
10. Completely factor $x^2 - 8x + 15$.
11. When $(x - 2y)^2$ is written as a trinomial, the result is______________.
12. Adults paid \$2.50 each and children paid \$.75 each for their tickets to a baseball game. If 940 tickets were sold and the receipts were \$1,825, there were_______ adult tickets and______ children's tickets sold.

ANSWERS TO TESTS

Form A

1. $-x + 5$	2. $a^3 + 5a^2 + 4a$	3. $(3y + 2)(y + 1)$
4. 1	5. distributive	6. $-10y^3 - 5y^2 + 5y$
7. $3x(x^2 + 4x - 1)$	8. $(x - 4)(x - 2)$	9. $c(a - 3b)(a + 3b)$
10. $2(2x + y)(x + 3y)$	11. $9x^2 + 6x + 1$	12. 13; 7

Form B

1. $-3x$	2. $b^3 - 2b^2 - 3b$	3. $(2x - 1)(x + 3)$
4. 2	5. $-9a^3 + 3a^2 - 3a$	6. $2a(a^2 - 4a + 1)$
7. distributive law	8. $(a + 3b)(a + b)$	9. $x(y - 2)(y + 2)$
10. $(x - 5)(x - 3)$	11. $x^2 - 4xy + 4y^2$	12. 640; 300

VOCABULARY, UNIT IV

The frame in which each word is introduced is shown in parentheses.

common factor (75)

distributive law (5)

factoring (61)

standard form of a polynomial (277)